Lecture Not
Mathematic

Edited by A. Dold and B. Eckmann

Subseries: Forschungsinstitut für Mathematik, ETH Zürich

1012

Shmuel Kantorovitz

Spectral Theory of Banach Space Operators

C^k-classification, abstract Volterra operators, similarity, spectrality, local spectral analysis.

Springer-Verlag
Berlin Heidelberg New York Tokyo 1983

Author

Shmuel Kantorovitz
Department of Mathematics, Bar-Ilan-University
Ramat-Gan, Israel

AMS Subject Classifications (1980): 47-02, 46H30, 47A60, 47A65, 47A55, 47D05, 47D10, 47D40, 47B47, 47A10

ISBN 3-540-12673-2 Springer-Verlag Berlin Heidelberg New York Tokyo
ISBN 0-387-12673-2 Springer-Verlag New York Heidelberg Berlin Tokyo

Printing and binding: Beltz Offsetdruck, Hemsbach/Bergstr.
2146/3140-543210

To Ita, Bracha, Peninah, Pinchas, and Ruth.

Table of Content

0. Introduction.

We may view selfadjoint operators in Hilbert space as the best understood properly infinite dimensional abstract operators. If we desire to recuperate some of their nice properties without the stringent selfadjointness hypothesis, we are led to a "non-selfadjoint theory" such as Dunford's theory of spectral operators [5; Part III] or Foias' theory of generalized spectral operators [9,4], to mention only a few, and it is not our purpose to describe here any one of these. Our basic concept, as in Foias' theory or distribution theory (as opposed to Dunford's), will be the operational calculus (and not the resolution of the identity). However there will be very little overlapping between [4] and the present exposition. Indeed, we shall go in an entirely different direction: starting in an abstract setting, we shall reduce the general situation to a very concrete one, and we shall then concentrate on various problems within this latter framework or its abstract lifting.

These notes are based on lectures given at various universities in 1981, and present in a unified (and often simplified) way results scattered through our papers since 1964.

We proceed now with a more specific description of the main features of this exposition.

Let K be a compact subset of the real line R, and denote by $H_R(K)$ the algebra of all complex functions which are "real analytic" in a (real) neighborhood of K, with pointwise operations and the usual topology. A basic algebra $A(K)$ is a topological algebra of complex functions defined in a (real) neighborhood of K, with pointwise operations, such that $H_R(K) \subset A(K)$ topologically. If A is a unital complex Banach algebra, an $A(K)$-operational calculus for $a \in A$ is a continuous representation $\tau: A(K) \to A$ carried by K (that is, $\tau(f) = 0$ whenever $f \in A(K)$ vanishes in a neighborhood of K), such that $\tau(t) = a$ (where t denotes the function $t \to t$ on R). When such an operational calculus exists, we say that a is of class $A(K)$. In that case, the spectrum $\sigma(a)$ of a is necessarily contained in K, and $\tau|H_R(K)$ coincides with the classical analytic operational calculus for a. This means that we are concerned with the latter's extension to wider basic algebras, for appropriate elements $a \in A$. For example, if A is the Banach algebra $B(H)$ of all bounded linear operators on the Hilbert space H, and $A(K) = C(K)$, the continuous functions in a neighborhood of K (or on K), then $T \in B(H)$ is of class $C(K)$ if and only if T is similar to a self-adjoint operator with spectrum in K. Since any bounded (linear) operator with spectrum in K is of class $H_R(K)$, we are really interested in intermediate topological algebras $A(K)$, contained topologically between $H_R(K)$ and $C(K)$, and in the study of the corresponding classes $[A(K)]$ of elements of class $A(K)$ (in A). A mild assumption on the topology of $A(K)$ shows that it suffices to consider the intermediate algebras $C^n(K)$ of all complex functions defined in a real neighborhood of K, and continuously differentiable there up to the order n (with the usual topology). More specifically, we have the following "first reduction" (§3): if $A(K)$ is a homogeneous normable basic algebra, and if $a \in [A(K)]$, then there exists $n \geq 2$ such that $a \in [C^n(K)]$.

A simplified (equivalent) way to consider C^n-operational calculi is to take the Banach algebra $C^n[\alpha,\beta]$ as the domain of the representation τ, where $[\alpha,\beta]$ is a closed interval containing 0, without loss of generality. For $n \geq 1$, an example of an operator of class $C^n[\alpha,\beta]$ (which is not of class $C^{n-1}[\alpha,\beta]$) is the operator T_n defined in $C[\alpha,\beta]$ by $T_n = M + nJ$, where $M\colon f(x) \to xf(x)$, and $J\colon f(x) \to \int_0^x f(t)\,dt$. Our "second reduction" (§4) consists in showing that T_n is a kind of universal model for elements of class C^n (in any Banach algebra A). Let L_a denote the "left multiplication by a" operator in A ($L_a x = ax$, $x \in A$). Then $a \in A$ is of class $C^n[\alpha,\beta]$ if and only if there exists a continuous linear map $U\colon C[\alpha,\beta] \to A$, normalized by the condition $U1 = a^n/n!$, and intertwining L_a and T_n: $L_aU = UT_n$. When this is the case, U is unique, and is related to the $C^n[\alpha,\beta]$-operational calculus for a by a kind of Taylor formula. This so-called "weak representation for a" on $C[\alpha,\beta]$ is a useful tool in producing more concrete characterization theorems for elements of class C^n, but its main conceptual importance is in bringing down the abstraction level to the initial C-case (what are all the possible maps $U\colon C[\alpha,\beta] \to A$?) and to the concrete operator T_n on $C[\alpha,\beta]$. In this exposition, the weak representation motivates our effort to extract the basic mechanism making $T_n = M + nJ$ into an operator of class C^n. Observing that J satisfies the commutation relation $[M,J] = J^2$ with the operator M of class C, we first study general spectral properties of $s + \zeta v$ ($\zeta \in C$) with $s, v \in A$ satisfying the so-called Volterra relation $[s,v] = v^2$. It is shown that if $\sigma(s) \subset R$, then $\sigma(s + kv) = \sigma(s)$ for all integers k, and if s is of class C^m, then $s + kv$ is of class $C^{m+|k|}$. This result is then generalized to "perturbations" $s + \zeta v$ with ζ complex (§6). Let $S,V \in B(X)$, where X is a Banach space. Suppose (Standing Hypothesis): $[S,V] = V^2$, and $V = V(1)$, where $\{V(\zeta);\ \zeta \in C^+\}$ is a regular semigroup of operators on X (cf. Definition 6.0) whose boundary group $\{V(i\eta);\ \eta \in R\}$ (cf. Theorem 6.1) satisfies $||V(i\eta)|| \leq Ke^{\nu|\eta|}$ with $\nu < \pi$. Set $T_\zeta = S + \zeta V$, and let m,k denote non-negative integers. In this general setting, we show that if S is of class C^m with real spectrum, then $\sigma(T_\zeta) = \sigma(S)$ for all complex ζ, and T_ζ is of class C^{m+k} in the strip $|\mathrm{Re}\,\zeta| \leq k$. If S is of class C, then T_ζ is of class C^k if and only if $|\mathrm{Re}\,\zeta| \leq k$. This latter result is shown to be applicable in particular to the operators $T_\zeta = M + \zeta J$ in $C[0,N]$ or $L^p(0,N)$ ($0 < N < \infty$, $1 < p < \infty$). Some needed tools from the theory of convolution operators are included in §7, and are then applied in §8 to prove the regularity of certain semigroups, and in particular, of the Riemann-Liouville semigroup $\{J^\zeta;\ \zeta \in C^+\}$, where

$$(J^\zeta f)(x) = \Gamma(\zeta)^{-1} \int_0^x (x - t)^{\zeta-1} f(t)\,dt,$$

acting in either $C[0,N]$ or $L^p(0,N)$ ($0 < N < \infty$, $1 < p < \infty$).

A by-product of the theory presented in §6 is the following similarity result (valid under the Standing Hypothesis when $||e^{itS}|| = O(1)$). For $\zeta, \omega \in C$, T_ζ and T_ω are similar if $\mathrm{Re}\,\zeta = \mathrm{Re}\,\omega$, and only if $|\mathrm{Re}\,\zeta| = |\mathrm{Re}\,\omega|$. In the concrete case where $T_\zeta = M + \zeta J$ in either $C[0,N]$ or $L^p(0,N)$ ($0 < N < \infty$, $1 < p < \infty$), this can be strengthened to: T_ζ and T_ω are similar if and only if $\mathrm{Re}\,\zeta = \mathrm{Re}\,\omega$.

In §9, we discuss various refinements of this theorem and of the preceding C^k-classification result. Let S,V satisfy the Standing Hypothesis (in particular, this forces V to be quasinilpotent). Suppose $f \in H(0)_r$ (that is, f is analytic at 0 and $f(0)$ is real). If S is of class C^m with real spectrum, it is shown that $S + f(V)$ is of class C^{m+k} for $|\mathrm{Re} f'(0)| \leq k$. In case $m = 0$, $S + f(V)$ is of class C^k if and only if $|\mathrm{Re} f'(0)| \leq k$. If $f,g \in H(0)_r$ are such that $f(0) = g(0)$ and $\mathrm{Re} f'(0) = \mathrm{Re} g'(0)$, then $S + f(V)$ is similar to $S + g(V)$, and conversely, if $\|e^{itS}\| = O(1)$ and $S + f(V)$ is similar to $S + g(V)$, then $f(0) = g(0)$ and $|\mathrm{Re} f'(0)| = |\mathrm{Re} g'(0)|$. In particular, on either $C[0,N]$ or $L^p(0,N)$ $(0 < N < \infty,\ 1 < p < \infty)$, and for $f,g \in H(0)_r$, $M + f(J)$ is of class C^k if and only if $|\mathrm{Re} f'(0)| \leq k$, and is similar to $M + g(J)$ if and only if $f(0) = g(0)$ and $\mathrm{Re} f'(0) = \mathrm{Re} g'(0)$.

Since M is trivially spectral of scalar type on $L^p(0,N)$, the latter result (with $g \equiv 0$) implies in particular that $M + \zeta J$ is spectral (of scalar type) if $\mathrm{Re}\,\zeta = 0$. A more general problem is to determine the values of ζ for which $M + \zeta J$ (or its abstraction $T_\zeta = S + \zeta V$) is spectral (not necessarily of scalar type), that is, possesses a Jordan decompostion $A + N$, with A and N commuting, $A = \int \lambda E(d\lambda)$ for some spectral measure E, and N quasinilpotent. One of the results proved in §10 is the following general theorem: if S,V satisfy the Standing Hypothesis, and S is a scalar type spectral operator with real spectrum, then $T_\zeta = S + \zeta V$ is spectral if and only if $\mathrm{Re}\,\zeta = 0$. This applies in particular to $M + \zeta J$ acting in $L^p(0,N)$, and illustrates the **richness** of the class of C^n-operators $(n > 0)$ as compared to the class of spectral operators within the family $\{T_\zeta = M + \zeta J;\ \zeta \in \mathbb{C}\}$: C^n-operators correspond to the strip $|\mathrm{Re}\,\zeta| \leq n$, while spectral operators correspond to the imaginary axis!

In §11, some of the results of Sections 6 and 9 are extended to the case when S is an unbounded operator with domain $D(S)$ in X. Such a situation arises for example when $S = M$ and $V = J_\varepsilon$: $f(x) \to \int_0^x e^{-\varepsilon(x-t)} f(t)\,dt$ $(\varepsilon > 0)$, both acting in $L^p(0,\infty)$. The Banach algebra methods used in the bounded case are not available. However, when iS generates a strongly continuous group of operators $S(\cdot)$, methods of the theory of semigroups step in instead, and we obtain adequate versions of the main results. With a slightly modified "Standing

Hypothesis", it is shown for example that $S + \zeta V$ is similar to $S + \zeta V$ if $\mathrm{Re}\zeta = \mathrm{Re}\omega$, and when $\|S(\cdot)\| = O(1)$, only if $|\mathrm{Re}\zeta| = |\mathrm{Re}\omega|$. There is also a natural way to define C^n-classes among group generators. If S is of class C^m, then $S + \zeta V$ is of class C^{m+k} for all ζ in the strip $|\mathrm{Re}\zeta| \leq k$. These results apply in particular to $M + \zeta J_\varepsilon$ acting in $L^p(0,\infty)$ $(1 < p < \infty;\ \varepsilon > 0)$. In §12, similarity is further discussed in various unbounded settings. We show for example that $M + \zeta J$ and $M + \omega J$, acting in $L^p(0,\infty)$ $(1 < p < \infty)$ with their maximal domain, are similar whenever ζ, ω are non-zero complex numbers with $\mathrm{Re}\zeta = \mathrm{Re}\omega$.

As previously observed, C^n-operators are rarely spectral. It is proved however in §13 that "singular" C^n-operators in reflexive Banach space are spectral.

Localized versions of the operational calculus, the spectral decomposition, and the Jordan canonical form, are obtain in §13-14 for arbitrary operators with real spectrum.

1. Operational Calculus

Let K be a compact subset of the complex plane $\mathbb{C}$, and denote by $H(K)$ the topological algebra of all complex functions f defined and analytic in some neighborhood Ω_f of K (depending on f). The operations are defined pointwise. A net $\{f_\alpha\} \in H(K)$ converges to zero if there exists a fixed neighborhood Ω of K in which all the functions f are analytic, and $f_\alpha \to 0$ uniformly on every compact subset of Ω. If A is a Banach algebra with identity 1 and $a \in A$ has spectrum $\sigma(a)$ contained in K, there exists a continuous representation τ of $H(K)$ on A (that is, an algebra homomorphism of $H(K)$ into A sending the constant function 1 to the identity $1 \in A$) such that $\tau(\lambda) = a$ (λ denotes here the function $\lambda \to \lambda$ on $\mathbb{C}$).

τ is unique, and may be expressed by means of the Riesz-Dunford integral

$$\tau(f) = \frac{1}{2\pi i} \int_\Gamma f(\lambda)(\lambda 1 - a)^{-1} d\lambda \qquad (f \in H(K))$$

where $\Gamma = \Gamma(\Omega_f, K)$ is a finite union of oriented closed Jordan curves such that K is contained in the union of the interiors (cf. [11], Theorem 5.2.5). The map τ is called the <u>analytic operational calculus for a</u>. We often write $f(a)$ instead of $\tau(f)$ when $f \in H(K)$.

For special elements $a \in A$, the analytic operational calculus may be extended to topological function algebras which properly contain $H(K)$.

1.1 <u>Definition</u>. A <u>basic algebra</u> $A(K)$ is a topological algebra of complex functions defined in a neighborhood of K, with pointwise operations, which contains $H(K)$ topologically. An <u>$A(K)$-operational calculus for $a \in A$</u> is a continuous representation $\tau : A(K) \to A$ carried by K (that is $\tau(f) = 0$ whenever $f \in A(K)$ is zero in a neighborhood of K), such that $\tau(\lambda) = a$.

When such an o.c. exists, one says that <u>a is of class $A(K)$</u>. The set of all the elements of class $A(K)$ in A is denoted by $[A(K)]_A$, or briefly, by $[A(K)]$.

Let J_K be the closed ideal of all $f \in A(K)$ which vanish in a neighborhood of K. Since τ is carried by K, it induces a representation on A of the quotient algebra $A(K)/J_K$ of "K-germs of $A(K)$-functions". For the sake of simplicity we prefer to consider the o.c. as a map of functions into A.

1.2 Proposition. If $a \in [A(K)]$, then $\sigma(a) \subset K$.

Proof. If $\mu \notin K$, the function $f_\mu(\lambda) = (\mu-\lambda)^{-1}$ belongs to $H(K)$, hence to $A(K)$. Let τ be an $A(K)$-o.c. for a. Since $(\mu-\lambda)f_\mu(\lambda) = 1$ in a neighborhood of K, we have

$$1 = \tau[(\mu-\lambda)f_\mu] = \tau(\mu-\lambda)\tau(f_\mu) = (\mu 1-a)\tau(f_\mu),$$

and similarly $\tau(f_\mu)(\mu 1-a) = 1$. This shows that $\mu \notin \sigma(a)$.

In particular, we have

$$[H(K)]_A = \{a \in A;\ \sigma(a) \subset K\}.$$

1.3 Proposition. If $a \in [A(K)]$ and τ is an $A(K)$-o.c. for a, then $\tau/H(K)$ coincides with the analytic o.c. for a.

Proof. Let $R(K)$ be the algebra of rational functions with poles outside K. Since τ is a representation such that $\tau(\lambda) = a$, it coincides with the analytic o.c. τ_{an} on $R(K)$. The Proposition follows from the fact that $R(K)$ is dense in $H(K)$ ([43; p.256]) and both τ and τ_{an} are continuous on $H(K)$ (since $H(K)$ is topologically contained in $A(K)$ and $\tau: A(K) \to A$ is continuous).

2. Examples

2.1 [C(K)]

For $K \subset C$ compact, let $C(K)$ denote the topological algebra of all complex functions f defined and continuous in a neighborhood Ω_f of K. The operations in $C(K)$ are defined pointwise. A net $\{f_\alpha\} \subset C(K)$ converges to zero if there is a fixed neighborhood Ω of K where all the f_α are continuous and $f_\alpha \to 0$ uniformly on every compact subset of Ω. By the Tietze extension theorem (cf. [43; p.385]), $C(K)$ can be identified topologically with the (usual) Banach algebra (also denoted $C(K)$) of all complex continuous functions on K, with the supremum norm $||f|| = \sup_K|f|$. Clearly $H(K) \subset C(K)$ topologically.

Let A be a B^*-algebra with identity, and let $a \in A$ be a normal element ($aa^* = a^*a$). The closed $*$-subalgebra generated by 1 and a is isometrically $*$-isomorphic to $C(\sigma(a))$, with a corresponding to λ (cf. [5; p.879]). Let τ_G: $C(\sigma(a)) \to A$ be such an isomorphism into A. For any compact $K \supset \sigma(a)$, we may define τ: $C(K) \to A$ by

$$\tau(f) = \tau_G(f|\sigma(a)) \qquad (f \in C(K)).$$

Clearly, τ is a $C(K)$-o.c. for a.

If $b \in A$ is similar to a normal element $a \in A$ with $\sigma(a) \subset K$, that is $b = qaq^{-1}$ for some non-singular $q \in A$, we let

$$\tau(f) = q\tau_a(f)q^{-1} \qquad (f \in C(K)),$$

where τ_a is the $C(K)$-o.c. for a defined above. τ is a $C(K)$-o.c. for b. Thus

$$[C(K)]_A \supset S_A(K),$$

where $S_A(K)$ is the set of all elements of A similar to normal elements, with spectrum contained in K.

Let $B(H)$ be the B^*-algebra of all bounded linear operators on the Hilbert space H.

2.1.1. Theorem. $[C(K)]_{B(H)} = S_{B(H)}(K)$.

Proof. By the preceding discussion, we must show that if $T \in B(H)$ is of class $C(K)$, then there exists a non-singular $Q \in B(H)$ such that $T = QNQ^{-1}$ for some normal operator N with spectrum in K.

Let τ be a $C(K)$-o.c. for T, and let $f_{(t,s)}(\lambda) = \exp i(t\mathrm{Re}\lambda + s\mathrm{Im}\lambda)$ for $(t,s) \in R^2$ $(\lambda \in C)$. By continuity of τ, there exists $M > 0$ such that $||\tau(f_{(t,s)})|| \leq M \sup_{\lambda \in K}|f_{(t,s)}(\lambda)| = M$; hence $\{\tau(f_{(t,s)}); (t,s) \in R^2\}$ is a uniformly continuous and uniformly bounded group of operators on H. It follows that there exists a non-singular operator Q such that $\{Q^{-1}\tau(f_{(t,s)})Q; (t,s) \in R^2\}$ is a group of unitary operators. In particular, $\{Q^{-1}\tau(f_{(t,o)})Q; t \in R\}$ and $\{Q^{-1}\tau(f_{(o,s)})Q; s \in R\}$ are one-parameter unitary groups. Their respective generators $U = Q^{-1}\tau(\mathrm{Re}\lambda)Q$ and $V = Q^{-1}\tau(\mathrm{Im}\lambda)Q$ are commuting self-adjoint operators, so that $N = U+iV$ is normal. Hence $T = \tau(\lambda) = \tau(\mathrm{Re}\lambda) + i\tau(\mathrm{Im}\lambda)$ $= QUQ^{-1} + iQVQ^{-1} = QNQ^{-1}$.

The operator Q in the above proof is constructed by using the Banach generalized limit (or "invariant mean") on the space $\mathbb{B}(R^2)$ of bounded complex functions on R^2 (cf. [6; 10.7]). Let G be any group for which there exists a normalized invariant mean m (for example, a solvable group, or in particular, an abelian group). Let $\{T_g; g \in G\}$ be a uniformly bounded representation of G on the Hilbert space H. For $x,y \in H$, the function

$$g \to (T_g x, T_g y)$$

belongs to $\mathbb{B}(G)$. Define

$$[x,y] = m(T_g x, T_g y).$$

The properties of m imply that $[\cdot,\cdot]$ is a hermitian bilinear form on $H \times H$, and

$$M^{-2}||x||^2 \leq [x,x] \leq M^2||x||^2 \qquad (x \in H)$$

where M is a positive upper bound for $\{||T_g||; g \in G\}$. Hence there exists a bounded strictly positive selfadjoint operator P on H such that

$$[x,y] = (Px,y) \qquad (x,y \in H).$$

Let Q be the positive square root of P. Then for all $u,v \in H$ and $h \in G$,

$$(QT_hQ^{-1}u,\ QT_hQ^{-1}v) =$$

$$= (PT_hQ^{-1}u,\ T_hQ^{-1}v)$$

$$= [T_hQ^{-1}u,\ T_hQ^{-1}v] = m(T_gT_hQ^{-1}u,\ T_gT_hQ^{-1}v)$$

$$= m(T_{gh}Q^{-1}u,\ T_{gh}Q^{-1}v)$$

$$= m(T_gQ^{-1}u,\ T_gQ^{-1}v) = [Q^{-1}u,\ Q^{-1}v]$$

$$= (Q^2Q^{-1}u,\ Q^{-1}v) = (u,v),$$

where we used the invariance of m. This shows that QT_hQ^{-1} is a unitary operator for each $h \in G$.

2.2 $[C^m(K)]$.

Let K be a compact subset of R. When this is the case, it is convenient to consider topological algebras $A(K)$ of complex functions f defined in a real neighborhood Ω_f of K. The assumption $H(K) \subset A(K)$ (topologically) is replaced by $H_R(K) \subset A(K)$, where $f \in H_R(K)$ if and only if there exists $\tilde{f} \in H(K)$ such that $f = \tilde{f}$ in Ω_f ($\tilde{f}$ is uniquely determined up to "equivalence modulo K", where "$f \sim g \bmod K$ in $H(K)$" means that $f = g$ in a complex neighborhood of K). A net $\{f_\alpha\} \subset H_R(K)$ converges to zero if $\{\tilde{f}_\alpha\} \to 0$ in $H(K)$. The analytic operational calculus for $a \in A$ is defined by

$$\tau_{an}(f) = \tau(\tilde{f}) \qquad (f \in H_R(K)),$$

where τ is the Riesz-Dunford o.c. for a (with $\sigma(a) \subset K$). Proposition 1.2 remains unchanged, while Proposition 1.3 should read: $\tau | H_R(K) = \tau_{an}$.

For $m = 0,1,2,\ldots$, let $C^m(K)$ denote the topological algebra of all complex functions f defined in a real neighborhood Ω_f of K, with continuous derivatives $f^{(j)}$ in Ω_f for $0 \leq j \leq m$. A net $\{f_\alpha\} \subset C^m(K)$ converges to zero if $\bigcap_\alpha \Omega_{f_\alpha}$ contains a neighborhood Ω of K, and $f_\alpha^{(j)} \to 0$ uniformly on every compact subset of Ω for $0 \leq j \leq m$. $C^\infty(K)$ is defined in a similar fashion. The topological inclusion $H_R(K) \subset C^m(K)$ $(0 \leq m \leq \infty)$ follows easily from Cauchy's formula. Thus $C^m(K)$ is a basic algebra. The corresponding classes $[C^m(K)]$ will turn out to have a distinguished position among the "general" classes $[A(K)]$ (see Section 3).

We consider now a concrete example of an operator of class $C^m(K)$, which has itself a distinguished position in the general theory (see Section 4).

Let K be a compact interval containing 0.

Let $M: \varphi(x) \to x\varphi(x)$ and

$$J: \varphi(x) \to \int_0^x \varphi(t)\,dt,$$

acting in the Banach space $C(K)$.

For $m = 0,1,2,\ldots$, set

$$T_m = M + mJ.$$

Clearly, $T_0 = M$ is of class $C(K)$. Its $C(K)$-o.c. is given by

$$\tau_0(f): \varphi(x) \to f(x)\varphi(x) \qquad (\varphi, f \in C(K)).$$

The range of J^m consists of all $\varphi \in C^m(K)$ such that $\varphi^{(j)}(0) = 0$ for $0 \leq j < m$, and is therefore an ideal in $C^m(K)$ (by Leibnitz' formula). Since J is one-to-one, J^{-m} is well-defined (with domain $D(J^{-m}) =$ range J^m), and acts as m-th order differentiation. It follows that the map $\tau_m: C^m(K) \to B(C(K))$ given by $\tau_m(f) = J^{-m}\tau_0(f)J^m$ $(f \in C^m(K))$ is a well-defined representation of $C^m(K)$ on $C(K)$. By Leibnitz' formula,

$$(*) \quad \tau_m(f) = \sum_{j=0}^{m} \binom{m}{j} \tau_0(f^{(j)})J^j \qquad (f \in C^m(K)).$$

This implies that τ_m is continuous, is carried by K, and $\tau_m(\lambda) = T_m$. Thus $T_m \in [C^m(K)]$ (in particular, $\sigma(T_m) \subset K$). In fact, $T_m \in [C^m(K)] \setminus [C^{m-1}(K)]$ $(m \geq 1)$. Indeed, a necessary condition for $a \in [C^r(K)]_A$ (for $0 \leq r < \infty$ fixed) is that

$$||e^{ita}|| = O(|t|^r) \quad \text{as} \quad |t| \to \infty.$$

To see this, let τ be a $C^r(K)$-o.c. for a, and $f_t(s) = e^{its}$ $(t,s \in R)$. By continuity of τ, there corresponds a positive constant M_Q to each compact subset Q of R such that

$$||e^{ita}|| = ||\tau(f_t)|| \leq M_Q \sum_{j=0}^{r} \max_{s \in Q} |f_t^{(j)}(s)|$$

$$= M_Q \sum_{j=0}^{r} |t|^j = O(|t|^r).$$

Now by (*)

$$e^{itT_m} = \tau_m(f_t) = \sum_{j=0}^{m} \binom{m}{j} (it)^j e^{itM} J^j .$$

Since $||e^{itM}|| = 1$, we have

$$t^{-m} e^{itT_m} \underset{|t|\to\infty}{\to} i^m e^{itM} J^m \qquad \text{in } B(C(K)).$$

Hence $|t|^{-m} ||e^{itT_m}|| \underset{|t|\to\infty}{\to} ||e^{itM} J^m|| = ||J^m|| \neq 0$, and so $||e^{itT_m}|| \neq 0(|t|^{m-1})$ as $|t| \to \infty$. More generally, it will be seen later that $T_\alpha = M + \alpha J$ $(\alpha \in C)$ acting in $C(K)$ or $L^p(K)$ $(1 < p < \infty)$ is of class $C^m(K)$ for a given m $(0 \leq m < \infty)$ if and only if $|\mathrm{Re}\,\alpha| \leq m$.

2.2 $A_n(K)$

Let A be a Banach algebra with unit, $n \in A$, and $K \subset R$ compact, Let $f \in C^\infty(K)$ (as in §2,2, it is understood that f is C^∞ in a real neighborhood Ω_f of K). For each compact subset Q of Ω_f, denote

$$||f||_{n,Q} = \sum_{j=0}^{\infty} \sup_Q |f^{(j)}| \; ||n^j||/j!$$

and

$$A_n(K) = \{f \in C^\infty(K);\ ||f||_{n,Q} < \infty \quad \text{for all compact subsets } Q \subset \Omega_f\}.$$

It follows from Leibnitz' formula that $A_n(K)$ is an algebra. We topologize it as follows: a net $\{f_\alpha\} \subset A_n(K)$ converges to zero if all f_α are C^∞ on a fixed neighborhood Ω of K, and $||f_\alpha||_{n,Q} \to 0$ for each compact subset Q of Ω.

If $f \in H_R(K)$ and $\tilde{f} \in H(K)$ coincides with f in Ω_f, then for each compact subset Q of Ω_f and for $\Gamma = \Gamma(\Omega_{\tilde{f}}, Q)$, Cauchy's formula implies that

$$|f^{(j)}(t)| \leq j! \int_\Gamma |\tilde{f}(\lambda)| \; |\lambda - t|^{-j-1} \; |d\lambda|/2\pi$$

$$\leq j! |\Gamma| \delta^{-j-1} (2\pi)^{-1} \sup_\Gamma |\tilde{f}| \qquad (t \in Q),$$

where $\delta = \mathrm{dist}(\Gamma, Q)$. Hence

$$||f||_{n,Q} \leq (2\pi)^{-1} |\Gamma| \sup_\Gamma |\tilde{f}| \sum_{j=0}^{\infty} ||n^j|| \delta^{-j-1} < \infty$$

provided that $\lim_{j\to\infty} (||n^j|| \delta^{-j-1})^{1/j} < 1$. However this limit equals $r(n)/\delta$, where $r(n)$ denotes the spectral radius of n. Thus, if n is quasi-nilpotent (that is, $r(n) = 0$), we have $||f||_{n,Q} < \infty$ for all $f \in H_R(K)$ and $Q \subset \Omega_f$, i.e., $H_R(K) \subset A_n(K)$. This inclusion is topological. Indeed, if $\{f_\alpha\} \subset H_R(K)$ converges to zero in $H_R(K)$, then all $\tilde{f}_\alpha$ are analytic in a fixed complex neighborhood Ω of K. Taking $\Gamma = \Gamma(\Omega, Q)$ in the preceding estimate (for each compact subset $Q \subset \Omega \cap R$), we have

$$||f_\alpha||_{n,Q} \le (2\pi)^{-1}|\Gamma| \sup_\Gamma |\tilde{f}_\alpha| \sum_{j=0}^{\infty} ||n^j||\delta^{-j-1} \to 0,$$

i.e. $f_\alpha \to 0$ in $A_n(K)$.

Thus, the classes $[A_n(K)]$ make sense (cf. Definition 1.1). Note that if n is nilpotent, then $A_n(K) = C^\infty(K)$ as sets, but the topology of $C^\infty(K)$ is stronger: the topology of $A_n(K)$ is equivalent to that of $C^r(K)$ (relativized to $A_n(K) = C^\infty(K)$), where r is the least non-negative integer such that $n^{r+1} = 0$.

Let $s \in [C(K)]_A$ and let n be a quasi-nilpotent element of A which commutes with s. Then $a = s+n \in [A_n(K)]_A$. Indeed, let τ_s be the $C(K)$-o.c. for s, and define

$$\tau_a(f) = \sum_{j=0}^{\infty} \tau_s(f^{(j)})n^j/j! \qquad (f \in A_n(K)). \tag{2.3.1}$$

Since $||\tau_s(h)|| \le M \sup_K |h|$ $(h \in C(K))$, it follows that the series above converges in A and $||\tau_a(f)|| \le M||f||_{n,K}$ $(f \in A_n(K))$. Thus τ_a is a continuous linear map of $A_n(K)$ into A. Clearly, τ_a is carried by K. Note that n commutes with $\tau_s(h)$ for any $h \in C(K)$ (because n commutes with polynomials in s, and polynomials are dense in $C(K)$). Therefore, if $f, g \in A_n(K)$,

$$\tau_a(fg) = \sum_{j=0}^{\infty} \sum_{m=0}^{j} \binom{j}{m} \tau_s(f^{(m)})\tau_s(g^{(j-m)})n^j/j!$$

$$= \sum_{m=0}^{\infty} \sum_{j=m}^{\infty} \tau_s(f^{(m)})(n^m/m!)\tau_s(g^{(j-m)})(n^{j-m}/(j-m)!)$$

$$= \tau_a(f)\tau_a(g).$$

Finally, $\tau_a(1) = 1$ and $\tau_a(\lambda) = \tau_s(\lambda) + \tau_s(1)n = s+n = a$. Thus τ_a is an $A_n(K)$-o.c. for a.

In case n is nilpotent $(n^{r+1} = 0)$, we have

$$\tau_a(f) = \sum_{j=0}^{r} \tau_s(f^{(j)})n^j/j! \qquad (f \in A_n(K) = C^\infty(K)). \tag{2.3.2}$$

This formula can be used to extend τ_a as a $C^r(K)$-operational calculus for a, that is $a \in [C^r(K)]$.

We summarize the above discussion:

2.3.1 Proposition. Let $s \in [C(K)]_A$ and let n be a quasi-nilpotent element of A which commutes with s. Then $a = s+n \in [A_n(K)]$, and an $A_n(K)$-o.c. for a is given by (2.3.1). If n is nilpotent $(n^{r+1} = 0)$, then $a \in [C^r(K)]$ and a $C^r(K)$-o.c. for a is given by (2.3.2)

3. First reduction

Let $H_R(R)$ denote the algebra of all complex functions f on R admitting an analytic continuation $\tilde{f}$ in some strip $|\text{Im}\,\lambda| < \delta_f$. A net $\{f_\alpha\}$ converges to zero in $H_R(R)$ if the f_α admit analytic continuations $\tilde{f}_\alpha$ in a fixed strip $|\text{Im}\,\lambda| < \delta$ and $\tilde{f}_\alpha \to 0$ uniformly on every compact subset of the strip. Recalling the conventions about $A(K)$ when K is a compact subset of R, we have $H_R(R) \subset A(K)$ topologically (cf. §2.2). For each $t \in R$, the map w_t of $H_R(R)$ into itself is defined by

$$(w_t f)(s) = f(ts) \qquad (s \in R;\ f \in H_R(R)).$$

Each w_t is a continuous linear map of $H_R(R)$ into itself. In case it remains continuous (for each $t \in R$) when $H_R(R)$ is equipped <u>with the relative $A(K)$-topology</u>, we say that $A(K)$ is <u>homogeneous</u>. The abstract situation reduces then to a very concrete one.

3.1 <u>Theorem.</u> Let $K \subset R$ be compact, and supose that the basic algebra $A(K)$ is homogeneous and normable. Then for any Banach algebra A,

$$[A(K)]_A \subset \bigcup_{n\geq 2} [C^n(K)]_A$$

<u>Proof</u>. Let $a \in [A(K)]_A$ and let $\tau : A(K) \to A$ be an $A(K)$-o.c. for a. Let $||\cdot||$ be a norm on $A(K)$ that induces its topology, and let $||\tau||$ be the corresponding norm of τ. For each $t \in R$, w_t is by assumption a bounded linear operator in the normed space $(H_R(R), ||\cdot||)$; denote its operator norm by $||w_t||$.

Let $f \in H_R(R)$ be the function $f: s \to e^{is}$. For $n = 1,2,\ldots$ and $t \in R$, we have

$$||e^{inta}|| = ||\tau(w_{nt} f)|| \leq ||\tau||\ ||w_t w_n f|| \leq ||\tau||\ ||w_t||\ ||w_n f||$$

Therefore the function

$$F(t) = \sup_n \frac{||e^{inta}||}{||w_n f||} \qquad (t \in R)$$

is finite valued and belongs to the first class of Baire (as the supremum of a sequence of continuous functions of t). Hence F has some point of continuity t_0 (cf. [11; p.765]). There exists therefore $\delta > 0$ such that

$$H = \sup \{F(t),\ |t-t_0| < \delta\} < \infty .$$

Now, for $|t-t_0| < \delta$ and $n = 1,2,\ldots$

$$||e^{in(t-t_0)a}|| \le ||e^{inta}||\ ||e^{-int_0a}||$$

$$\le F(t)F(-t_0)||w_nf||^2 \le C||w_nf||^2$$

with $C = HF(-t_0)$. That is,

$$||e^{insa}|| \le C||w_nf||^2 \qquad (3.1.1)$$

for $s \in R$, $|s| < \delta$, and all $n = 1,2,\ldots$.

For $t \in R$ fixed, let k be the first integer such that $2^{-k}|t| < \delta$. Since $2^{-k+1}|t| \ge \delta$, we have

$$k \le c + \log_2|t| \qquad (c = 1 - \log_2\delta) \qquad (3.1.2)$$

Take $s = 2^{-k}t$ and $n = 2^k$ in (3.1.1). Thus

$$||e^{ita}|| \le C||w_{2^k}f||^2 \le C(||w_2||^k||f||)^2.$$

Since $w_t1 = 1$, we have $||w_t|| \ge 1$ for any t. Hence, by (3.1.2),

$$||e^{ita}|| \le C_1|t|^h \qquad (t \in R)$$

where $C_1 = C||w_2||^{2c}||f||^2$ and $h = 2\log_2||w_2|| \ge 0$.

The theorem follows now from the following

3.2 Lemma. Let $a \in A$ be such that

$$||e^{ita}|| = O(|t|^h) \qquad \text{as} \quad |t| \to \infty$$

for some $h \ge 0$. Then $a \in [C^n(K)]$ for $n \ge [h]+2$ and any compact set $K \subset R$ containing $\sigma(a)$.

Proof. Observe first that the growth condition of the hypothesis implies that $\sigma(a) \subset R$. More generally, since the function $\log||e^{ita}||$ is subadditive on R, the limits

$$s^{\pm}(a) = \lim_{t\to\pm\infty} t^{-1}\log||e^{-ita}|| \qquad (3.2.1)$$

exist (cf. [11; p.244]). By the spectral radius formula and the spectral mapping theorem, we obtain

$$s^{+}(a) = \lim_{n\to\infty} n^{-1}\log\|e^{-ina}\| = \log \lim_{n\to\infty} \|(e^{-ia})^n\|^{1/n}$$

$$= \log \sup\{|\lambda|,\ \lambda \in \sigma(e^{-ia})\}$$

$$= \sup \log\{|e^{-i(\xi+i\eta)}|;\ \xi+i\eta \in \sigma(a)\}$$

$$= \sup \{\eta;\ \xi+i\eta \in \sigma(a)\} \ = \sup \operatorname{Im} \sigma(a).$$

Similarly, $s^{-}(a) = \operatorname{Inf} \operatorname{Im} \sigma(a)$.

In particular, $\sigma(a)$ is real if and only if

$$\lim_{|t|\to\infty} t^{-1} \log\|e^{ita}\| = 0. \tag{3.2.2}$$

This is certainly the case when $\|e^{ita}\| = O(|t|^h)$, as we observed above.

Fix $n \geq [h] + 2$. For $f \in C_0^{\,n}(R)$ (C^n-function on R with compact support), denote by $\hat{f}$ the Fourier transform of f,

$$\hat{f}(t) = \frac{1}{2\pi}\int_R e^{-its} f(s)ds \qquad (t \in R) .$$

Since $\hat{f}$ is $O(|t|^{-n})$ as $|t| \to \infty$ (integrate by parts n times!), we have $\|\hat{f}(t)e^{ita}\| = O(|t|^{-1-\varepsilon})$ as $|t| \to \infty$ with $\varepsilon \geq 1 - (h-[h]) > 0$. Let

$$\tau(f) = \int_R \hat{f}(t)e^{ita}dt, \qquad f \in C_0^{\,n}(R). \tag{3.2.3}$$

By the preceding remark, the integral converges absolutely in A.

If $f,g \in C_0^{\,n}(R)$,

$$\tau(f)\tau(g) = \int_R \hat{f}(t)e^{ita}dt \int_R \hat{g}(s)e^{isa}ds$$

$$= \int_{R^2}\hat{f}(t)\hat{g}(s)e^{i(t+s)a}dtds$$

$$= \int_R(\int_R\hat{f}(u-s)\hat{g}(s)ds)e^{iua}du$$

$$= \int_R(fg)\hat{}\,(u)e^{iua}du = \tau(fg).$$

By the Lebesgue Dominated Convergence Theorem (for vector functions, cf. [5; p.124]),

$$\tau(f) = \lim_{\varepsilon\to 0+} \int_R\hat{f}(t)e^{-\varepsilon|t|}e^{ita}dt$$

$$= \lim_{\varepsilon\to 0+} \int_R f(s)(e^{-\varepsilon|t|}e^{ita})\hat{}\,(s)ds$$

However, by [11; Theorem 11.2.1],

$$2\pi(e^{-\varepsilon|t|}e^{ita})^{\wedge}(s) = \int_0^\infty e^{-(\varepsilon+is)t}e^{ita}dt + \int_0^\infty e^{-(\varepsilon-is)t}e^{-ita}dt$$

$$= R(\varepsilon+is;ia) + R(\varepsilon-is;-ia)$$

$$= i(R(s+i\varepsilon;a) - R(s-i\varepsilon;a)),$$

where $R(\lambda;a) = (\lambda 1-a)^{-1}$ is the resolvent of a. Hence, for $f \in C_0^n(R)$,

$$\tau(f) = \lim_{\varepsilon\to 0+} \frac{1}{2\pi i}\int_R f(s)[R(s-i\varepsilon;a) - R(s+i\varepsilon;a)]ds \,. \tag{3.2.4}$$

If $f \in C_0^n(R)$ vanishes in a neighborhood of $K \supset \sigma(a)$, then $R(s;a)$ is continuous on supp f (the support of f); let M be the supremum of $||R(s;a)||$ on supp f. By the First Resolvent Equation, we have for $\varepsilon^2 < 1/2M^2$ and $s \in$ supp f

$$||R(s-i\varepsilon;a) - R(s+i\varepsilon;a)|| = 2\varepsilon||[(s-a)^2 + \varepsilon^2]^{-1}||$$

$$\leq \frac{2\varepsilon||R(s;a)||^2}{1-\varepsilon^2||R(s;a)||^2} \leq \frac{2\varepsilon M^2}{1-\varepsilon^2M^2} < 4\varepsilon M^2.$$

Therefore the integral in (3.2.4) has norm less than $(2/\pi)\varepsilon M^2\sup|f|\ |\text{supp } f|$ (for ε as above), hence $\tau(f) = 0$.

Now, for $f \in C^n(K)$, let $\tilde{f}$ be any function in $C_0^n(R)$ which coincides with f in a neighborhood of K, and set $\tau(f) = \tau(\tilde{f})$. By the preceding remark, $\tau(f)$ is well-defined. Clearly, the map $\tau: C^n(K) \to A$ is linear and multiplicative and is carried by K. If $f_\alpha \in C^n(K)$ converge to zero, there exists a fixed neighborhood Ω of K in which all f_α are C^n functions and $f_\alpha^{(j)} \to 0$ uniformly on every compact subset of $\Omega (0 \leq j \leq n)$. Let $\varphi \in C_0^n(R)$ have support $K' \subset \Omega$ and be equal to 1 in a neighborhood of K. Let $(\beta,\gamma) \supset K'$. By (3.2.3) integrations by parts, and Leibnitz' formula, we obtain

$$||\tau(f_\alpha)|| = ||\tau(\varphi f_\alpha)|| \leq \text{Const}\sum_{j=0}^n \max_{[\beta,\gamma]} |(\varphi f_\alpha)^{(j)}|$$

$$\leq \text{Const}\sum_{j=0}^n \max_{K'}|f_\alpha^{(j)}| \to 0.$$

Thus $\tau: C^n(K) \to A$ is continuous.

Estimates as those following (3.2.4) show that whenever $(\beta,\gamma) \supset \sigma(a)$, we have

$$\tau(f) = \lim_{\varepsilon\to 0+} \frac{1}{2\pi i}\int_\beta^\gamma f(s)[R(s-i\varepsilon;a) - R(s+i\varepsilon;a)]ds \tag{3.2.5}$$

for $f \in C_0^n(R)$.

In particular, taking $f \in C_0^n(R)$ equal to 1 in (β,γ) we see that if $\sigma(a) \subset (\beta,\gamma)$,

$$\tau(1) = \lim_{\varepsilon\to 0+} \frac{1}{2\pi i} \int_\beta^\gamma R(s-i\varepsilon;a) - R(s+i\varepsilon;a)\, ds.$$

Let Γ_ε be the rectangle with vertices $\beta \pm i\varepsilon$ and $\gamma \pm i\varepsilon$ (described in the positive direction). By continuity of the resolvent $R(\lambda;a)$ on the resolvent set $\rho(a)$, its integrals over the vertical sides of Γ_ε have limit zero when $\varepsilon \to 0+$. Hence

$$\tau(1) = \lim_{\varepsilon\to 0+} \frac{1}{2\pi i} \int_{\Gamma_\varepsilon} R(\lambda;a)\, d\lambda = 1.$$

Similarly,

$$\begin{aligned} a &= \lim_{\varepsilon\to 0+} \frac{1}{2\pi i} \int_{\Gamma_\varepsilon} \lambda R(\lambda;a)\, d\lambda \\ &= \lim_{\varepsilon\to 0+} \Big\{\frac{1}{2\pi i} \int_\beta^\gamma s[R(s-i\varepsilon;a) - R(s+i\varepsilon;a)]ds - \frac{\varepsilon}{2\pi} \int_\beta^\gamma [R(s-i\varepsilon;a) + R(s+i\varepsilon;a)]ds\Big\}. \end{aligned}$$

Since $\sigma(a) \subset R$, we have

$$\int_\beta^\gamma R(s+i\varepsilon;a)ds = \int_1^\varepsilon R(\gamma+it;a)dt + \int_\beta^\gamma R(s+i;a)ds + \int_\varepsilon^1 R(\beta+it;a)dt \to z^+ \in A$$

as $\varepsilon \to 0+$, and similarly

$$\int_\beta^\gamma R(s-i\varepsilon;a)ds \to z^- \in A \quad \text{as} \quad \varepsilon \to 0+.$$

Hence, taking $f \in C_0^n(R)$ such that $f(s) = s$ in (β,γ), (3.2.5) implies that

$$\tau(s) = \lim_{\varepsilon\to 0+} \frac{1}{2\pi i} \int_\beta^\gamma s[R(s-i\varepsilon;a) - R(s+i\varepsilon;a)]ds = a.$$

Thus τ is a $C^n(K)$-o.c. for a, and Lemma 3.2 is proved.

Theorem 3.1 reduces the abstract situation to the concrete C^n-case: if $a \in [A(K)]$ for a homogeneous normable algebra $A(K)$, then $a \in [C^n(K)]$ for some $n \geq 2$. This result also stresses the distinguished position of the classes $[C^n(K)]$ in the operational calculus theory.

In a suitable sense, the study of $C^n(K)$-operational calculi can be reduced further to that of certain continuous linear maps $U: C(K) \to A$. This will be done in the next section.

4. Second reduction

For our present purpose, it will be convenient to denote by $C^n[\alpha,\beta]$ the Banach algebra of all complex C^n-functions in the closed interval $[\alpha,\beta]$ ($\alpha < \beta$; $n \geq 0$ integer), with pointwise operations and norm

$$||f|| = \sum_{j=0}^{n} \sup_{[\alpha,\beta]} |f^{(j)}|/j! \quad .$$

This should not be confused with the topological algebra $C^n([\alpha,\beta])$ of Section 2.2.

If A is any locally convex topological algebra with identity over C, a $C^n[\alpha,\beta]$-operational calculus for $a \in A$ is a continuous representation τ of $C^n[\alpha,\beta]$ in A such that $\tau(\lambda) = a$. Since polynomials are dense in $C^n[\alpha,\beta]$ such a map τ is unique if it exists. As before, $a \in A$ is of class $C^n[\alpha,\beta]$ if there exists a $C^n[\alpha,\beta]$-o.c. for a, and $[C^n[\alpha,\beta]]_A$ stands for the set of all such elements in A (which are also called $C^n[\alpha,\beta]$-elements of A).

If K is any compact subset of (α,β), we have $C^n[\alpha,\beta] \subset C^n(K)$ topologically. Therefore, if τ is a $C^n(K)$-o.c. for a, then $\tau|C^n[\alpha,\beta]$ is a $C^n[\alpha,\beta]$-o.c. for a carried by K. Conversely, if τ_0 is a $C^n[\alpha,\beta]$-o.c. for a carried by K, and $f \in C^n(K)$, pick $\tilde{f} \in C^n[\alpha,\beta]$ which coincides with f in a neighborhood of K, and define $\tau(f) = \tau_0(\tilde{f})$. One verifies easily that τ is a $C^n(K)$-o.c. for a carried by K (see argument in proof of Lemma 3.2). Thus $a \in [C^n(K)]$ if and only if $a \in [C^n[\alpha,\beta]]$. Since this is true for any interval $[\alpha,\beta]$ such that $K \subset (\alpha,\beta)$, we may (and shall) assume that $0 \in [\alpha,\beta]$.

Let $T_n = M+nJ$ be the $C^n[\alpha,\beta]$-operator in $C[\alpha,\beta]$ discussed in Section 2.2. Denote by L_a the "left multiplication by a" operator in A ($L_a x = ax$, $x \in A$). We may now state our "second reduction":

4.1 Theorem. Let A be a locally convex algebra with identity, $[\alpha,\beta]$ an interval containing 0, and n a non-negative integer. Then $a \in A$ is of class $C^n[\alpha,\beta]$ if and only if there exists a continuous linear map $U: C[\alpha,\beta] \to A$ such that

$$U1 = a^n/n! \qquad (4.1.1)$$

and .

$$L_a U = UT_n \tag{4.1.2}$$

When this is the case, U is unique, and is related to the $C^n[\alpha,\beta]$-operational calculus τ for a by the identities

$$U = \tau J^n \tag{4.1.3}$$

$$\tau(f) = \sum_{0\le j\le n-1} f^{(j)}(0)a^j/j! + Uf^{(n)} \tag{4.1.4}$$

$(f \in C^n[\alpha,\beta])$.

<u>Proof</u>. Let $a \in [C^n[\alpha,\beta]]_A$; let τ be the $C^n[\alpha,\beta]$-o.c. for a, and define U by (4.1.3). Clearly $U: C[\alpha,\beta] \to A$ is linear and continuous, $U1 = \tau(\lambda^n/n!) = a^n/n!$, and since

$$MJ^n = J^n T_n, \tag{4.1.5}$$

we have for any $h \in C[\alpha,\beta]$

$$\begin{aligned} L_a Uh &= a\tau(J^n h) = \tau(\lambda)\tau(J^n h) \\ &= \tau(MJ^n h) = \tau(J^n T_n h) = UT_n h. \end{aligned}$$

Conversely, suppose there exists $U: C[\alpha,\beta] \to A$ continuous and linear, satisfying (4.1.1) and (4.1.2). Define τ by (4.1.4). Then τ is a continuous linear map of $C^n[\alpha,\beta]$ into A, which satisfies (4.1.3) and

$$\tau(\lambda^m) = a^m \tag{4.1.6}$$

for $m \le n$ (cf. (4.1.1)).

Suppose (4.1.6) is valid for some $m \ge n$. By (4.1.3) and (4.1.2),

$$\begin{aligned} \tau(\lambda^{m+1}) &= \frac{(m+1)!}{(m-n+1)!}\tau(J^n\lambda^{m-n+1}) = \frac{(m+1)!}{(m-n+1)!}U(\lambda^{m-n+1}) \\ &= \frac{m!}{(m-n)!}UT_n(\lambda^{m-n}) = \frac{m!}{(m-n)!}L_a U(\lambda^{m-n}) \\ &= aU((\lambda^m)^{(n)}) = a\tau(\lambda^m) = a^{m+1} \end{aligned}$$

It follows by induction that (4.1.6) is valid for all $m = 0,1,2,\ldots$. Since the polynomials are dense in $C^n[\alpha,\beta]$, it follows that τ is multiplicative, that is, $a \in [C^n[\alpha,\beta]]$ and τ is the $C^n[\alpha,\beta]$-o.c. for a. Since τ is unique and (4.1.3) is necessary, the uniqueness of U follows.

The unique map $U: C[\alpha,\beta] \to A$ intertwining L_a and T_n (relation (4.1.2)) is called the "weak" representation of a on $C[\alpha,\beta]$. Since U is generally not one-to-one and onto, the "weak" canonical model T_n for $a \in [C^n[\alpha,\beta]]$ cannot be expected to preserve properties of a. Yet, the "weak representation theorem" 4.1 is a useful tool for obtaining characterizations of elements in $[C^n[\alpha,\beta]]$.

Let P denote the algebra of all polynomials over C (restricted to real variable). Write $\Delta = [\alpha,\beta]$, $||f||_\Delta = \max_\Delta |f|$, and set

$$||a||_{n,\Delta} = \sup_{0 \neq p \in P} ||(J^n p)(a)||/||p||_\Delta \qquad (a \in A)$$

where A is a Banach algebra with identity. Since $J^n p \in P$ for $p \in P$, $||a||_{n,\Delta}$ makes sense.

4.2 Corollary. $a \in [C^n[\alpha,\beta]]_A$ if and only if $||a||_{n,\Delta} < \infty$. When this is the case, $||a||_{n,\Delta} = ||U||$, where $U: C[\alpha,\beta] \to A$ is the weak representation of a on $C[\alpha,\beta]$.

Proof. If $a \in [C^n[\alpha,\beta]]$, let τ and U be respectively the $C^n[\alpha,\beta]$-o.c. and the weak representation of a. Then by (4.1.3),

$$||a||_{n,\Delta} = \sup_{0 \neq p \in P} ||\tau(J^n p)||/||p||_\Delta$$

$$= \sup ||Up||/||p||_\Delta = ||U|| < \infty$$

Conversely, if $||a||_{n,\Delta} < \infty$, consider P as a (dense) linear manifold in the Banach space $C[\alpha,\beta]$ and define $U_0: P \to A$ by

$$U_0 p = (J^n p)(a) = \tau_{an}(J^n p), \qquad p \in P,$$

where τ_{an} denotes the analytic operational calculus for a. U_0 is a continuous linear map, and has a unique extension as a continuous linear map $U: C[\alpha,\beta] \to A$ with norm $||U|| = ||U_0|| = ||a||_{n,\Delta}$. Clearly $U1 = U_0 1 = a^n/n!$. Since P is T_n-invariant, the calculation in the first part of the proof of Theorem 4.1 (with τ_{an} replacing τ) shows that $L_a U_0 = U_0 T_n$ on P, and therefore $L_a U = UT_n$ by continuity. Hence $a \in [C^n[\alpha,\beta]]$ by Theorem 4.1.

Let us say that a Banach algebra element a is of class C^n (or is a C^n-element) if $a \in [C^n[\alpha,\beta]]_A$ for some closed interval $\Delta = [\alpha,\beta]$. This is equivalent to saying that there exists a continuous representation

$\tau: C^n(R) \to A$ such that $\tau(\lambda) = a$. We call such a map a $C^n(R)$-o.c. for a.
Without loss of generality, we may always assume that $0 \in \Delta$ (cf. beginning of §4).
We are interested in a criterion for $a \in A$ to be of class C^n, with no reference to "some" undetermined interval Δ. Set

$$E_n(t;a) = \sum_{k=0}^{\infty} (it)^k a^{k+n}/(k+n)!\,, \qquad t \in R$$

$$\Omega = \{(\underline{c},\underline{t}) \in C^m \times R^m;\ \sup_{s \in R} \Big|\sum_{k=1}^{m} c_k \exp(it_k s)\Big| \le 1,\quad m = 1,2,\dots\},$$

and

$$v_n(a) = \sup_{(\underline{c},\underline{t}) \in \Omega} \Big|\Big|\sum_{k=1}^{m} c_k E_n(t_k;a)\Big|\Big|.$$

4.3 Theorem. $a \in A$ is of class C^n if and only if $v_n(a) < \infty$.

Proof. Necessity. Let $a \in A$ be of class C^n, and let τ and U be the $C^n[\alpha,\beta]$-operational calculus and the weak representation on $C[\alpha,\beta]$ for a. For each $t \in R$, set $f_t(s) = e^{its}$. Then

$$(J^n f_t)(s) = \sum_{k=0}^{\infty} (it)^k s^{k+n}/(k+n)!$$

and therefore, by (4.1.3)

$$Uf_t = E_n(t;a), \quad t \in R \tag{4.3.1}$$

Hence, for $(\underline{c},\underline{t}) \in \Omega$,

$$\Big|\Big|\sum_{k=1}^{m} c_k E_n(t_k;a)\Big|\Big| = \Big|\Big|U\Big(\sum_{k=1}^{m} c_k f_{t_k}\Big)\Big|\Big| \le ||U||,$$

that is, $v_n(a) \le ||U||$.

Sufficiency. Let A_1^* denote the closed unit ball of the dual A^* of A. Assuming $v_n(a) < \infty$, we have for each $a^* \in A_1^*$

$$\sup_{(\underline{c},\underline{t}) \in \Omega} \Big|\sum_{k=1}^{m} c_k a^* E_n(t_k;a)\Big| \le v_n(a).$$

By the Bochner criterion (cf. [2] or [44; p.32]), there exists a finite regular complex Borel measure $\mu(\cdot|a^*)$ such that

$$||\mu(\cdot|a^*)|| \le v_n(a) \tag{4.3.2}$$

and for all $t \in R$

$$a^* E_n(t;a) = \int_R e^{its} \mu(ds|a^*) \tag{4.3.3}$$

Let

$$E_n(z;a) = \sum_{k=0}^{\infty} (iz)^k a^{k+n}/(k+n)! \qquad z \in C.$$

Since $||E_n(z;a)|| \le ||a^n|| \exp(||a|| \, |z|)/n!$, the entire function $g(z) = a^* E_n(z;a)$ is of exponential type $\le ||a||$, and is bounded on R by $v_n(a)$.

By the Paley-Wiener-Schwartz theorem applied to g considered as a tempered distribution on R (cf [10; p.145]), the distribution Fourier transform $\hat{g} = \mu(\cdot|a^*)$ (cf. (4.3.3)) is supported by the interval $\Delta = [-||a||, ||a||]$.

Hence

$$a^* E_n(t;a) = \int_\Delta e^{its} \mu(ds|a^*), \qquad t \in R.$$

Expanding both sides in powers of t, we obtain

$$\int_\Delta s^k \mu(ds|a^*) = a^*\left(\frac{k!}{(k+n)!} a^{k+n}\right) = a^* \tau_{an}(J^n s^k)$$

for $k = 0,1,2,\ldots$, and so, by linearity,

$$a^* \tau_{an}(J^n p) = \int_\Delta p(s) \mu(ds|a^*), \qquad p \in P,$$

for all $a^* \in A_1^*$.

In particular, by (4.3.2),

$$||(J^n p)(a)|| = \sup\{|a^* \tau_{an}(J^n p)|;\ a^* \in A_1^*\} \le v_n(a) \sup|p|.$$

Hence $||a||_{n,\Delta} \le v_n(a)$, so that $a \in [C^n(\Delta)]$ by Corollary 4.2, and $||U|| \le v_n(a)$, where U is the weak representation of a on $C(\Delta)$ (actually, $||U|| = v_n(a)$ for $\Delta = [-||a||, ||a||]$; cf. "Necessity" part).

4.4 Proposition. Let $a \in A$ be of class C^n and let τ be a $C^n(R)$-o.c. for a. Then τ is carried by $\sigma(a)$.

Proof. Suppose $f \in C^n(R)$ vanishes in a neighborhood of $\sigma(a)$. Let $h_\lambda(t) = (\lambda - t)^{-1} f(t)$ $(\lambda \notin \operatorname{supp} f)$. Then $\tau(h_\lambda)$ is analytic on $(\operatorname{supp} f)'$ and $(\lambda 1 - a)\tau(h_\lambda) = \tau((\lambda - t)h_\lambda(t)) = \tau(f)$. Hence $\tau(h_\lambda) = (\lambda 1 - a)^{-1}\tau(f)$ is also analytic on $\rho(a) \supset \operatorname{supp} f$, that is, $\tau(h_\lambda)$ is entire. Clearly $\tau(h_\lambda) \to 0$ as $\lambda \to \infty$. By Liouville's theorem, $\tau(h_\lambda) \equiv 0$ and therefore $\tau(f) = 0$.

4.5 Corollary. Let $a \in A$ be of class C^n. Then $a \in [C^n(\sigma(a))]$.

5. Volterra elements

The weak representation theorem (Theorem 4.1) gives a special role to the operators $T_n = M+nJ$ $(n = 0,1,2,..)$ acting on $C[\alpha,\beta]$ $(0 \in [\alpha,\beta])$. We shall put these operators in perspective by deducing some of their properties from the commutation relation obviously satisfied by the pair (M,J):

$$[M,J] = J^2 \tag{5.1}$$

Here, $[\ ,\ ]$ denotes the Lie product

$$[a,b] = ab-ba$$

for a,b belonging to some ring. Since J is usually called "the Volterra operator", we shall refer to (5.1) as "the Volterra commutation relation".

Let A be an algebra with identity 1 over a field K. Polynomials over K operate on A in the usual way: $f(x) = \sum\lambda_k x^k$ for $x \in A$ and $f(\xi) = \sum\lambda_k \xi^k$. Denote by f' the formal derivative of the polynomial f. A derivation in A is a linear map $D: A \to A$ such that $D(ab) = (Da)b + a(Db)$ for all $a,b \in A$. For example, each $s \in A$ induces the "inner" derivation D_s, where

$$D_s a = [s,a].$$

5.1 Lemma (the Chain rule). Let D be a derivation in A, and suppose $v \in A$ commutes with Dv. Then

$$Df(v) = f'(v)Dv$$

for every polynomial f over K.

Proof. Since $D1 = 0$ is true for any derivation D, the lemma will follow by linearity from

$$Dv^k = kv^{k-1}Dv \qquad k = 1,2,... \tag{5.1.1}$$

We verify (5.1.1) by induction. It is trivial for $k = 1$. Assuming (5.1.1) for some k, we then have

$$Dv^{k+1} = (Dv^k)v + v^kDv = kv^{k-1}(Dv)v + v^kDv = (k+1)v^kDv$$

since Dv commutes with v.

5.2 Lemma. Let A be a Banach algebra with identity 1 (over $\mathbb{C}$), and let D be a continuous derivation in A. Suppose $v \in A$ commutes with Dv. Then the chain rule

$$Df(v) = f'(v)Dv$$

is valid for all $f \in H(\sigma(v))$.

Proof. If $a \in A$ is invertible,

$$0 = D1 = D(aa^{-1}) = (Da)a^{-1} + aDa^{-1}, \text{ and thus}$$

$Da^{-1} = -a^{-1}(Da)a^{-1}$. Hence, for $\lambda \in \rho(v)$,

$$D(\lambda 1-v)^{-1} = (\lambda 1-v)^{-1}Dv(\lambda 1-v)^{-1} = (\lambda 1-v)^{-2}Dv.$$

If $f \in H(\sigma(v))$ and $\Gamma = \Gamma(\Omega_f, \sigma(v))$ (cf. §1), the continuity of D implies that

$$Df(v) = \int_\Gamma f(\lambda) D(\lambda 1-v)^{-1} d\lambda/2\pi i$$
$$= \int_\Gamma f(\lambda)(\lambda 1-v)^{-2} d\lambda/2\pi i \; Dv = f'(v)Dv.$$

5.3 Notation. Let A be an algebra over a field K. For $a,b \in A$, we set

$$L_a x = ax;\; R_b x = xb;\; C(a,b) = L_a - R_b \quad (x \in A).$$

In particular, $C(a,a) = D_a$ (the inner derivation induced by a). Trivially, the left and right multiplication operators L_a and R_b commute. When A is a Banach algebra over C, the operator R_b should not be confused with the resolvent of $b \in A$, denoted $R(\xi,b)$ in the usual way $(\xi \in C)$. If a formal series $s = \sum a_j$ $(a_j \in A)$ converges in A, we use the notation s also for the sum of the series as an element of A.

Since $C(a,b)^j 1 = \sum_{k=0}^{j} (-1)^k \binom{j}{k} a^{j-k} b^k$,

we have, in case a commutes with b,

$$C(a,b)^j 1 = (a-b)^j.$$

5.4 Lemma. Let A be a Banach algebra with identity 1 over C; let $a,b \in A$ and $\xi \in \rho(a)$. Consider the formal series

$$b_L(\xi) = \sum_{j=0}^{\infty} (-1)^j R(\xi;a)^{j+1} \cdot C(a,b)^j 1 \tag{5.4.1}$$

$$b_R(\xi) = \sum_{j=0}^{\infty} C(b,a)^j 1 \cdot R(\xi;a)^{j+1} \tag{5.4.2}$$

Then if $b_L(\xi)$ $(b_R(\xi))$ converges in A, it is a left (right) inverse for $\xi 1-b$. In particular, if $\xi \in \rho(a)$ is such that both $b_L(\xi)$ and $b_R(\xi)$ converge in A, then $\xi \in \rho(b)$ and

$$R(\xi;b) = b_L(\xi) = b_R(\xi).$$

Proof. Suppose $b_L(\xi)$ converges in A for some $\xi \in \rho(a)$. Since $aR(\xi;a) = \xi R(\xi;a)-1$ and $C(a,b)$ commutes with $L_{R(\xi;a)}$, we have

$$b_L(\xi)b = R_b b_L(\xi) = [L_a - C(a,b)]b_L(\xi)$$

$$= \sum_{j=0}^{\infty}(-1)^j aR(\xi;a)^{j+1}\cdot C(a,b)^j 1 - \sum_{j=0}^{\infty}(-1)^j R(\xi;a)^{j+1}\cdot C(a,b)^{j+1} 1$$

$$= \xi b_L(\xi) - \sum_{j=0}^{\infty}(-1)^j R(\xi;a)^j\cdot C(a,b)^j 1 + \sum_{j=0}^{\infty}(-1)^{j+1}R(\xi;a)^{j+1}\cdot C(a,b)^{j+1} 1$$

$$= \xi b_L(\xi) - 1 .$$

Hence $b_L(\xi)(\xi 1-b) = 1$.

A similar computation, starting with

$$bb_R(\xi) = L_b b_R(\xi) = [R_a + C(b,a)]b_R(\xi),$$

shows that $b_R(\xi)$ is a right inverse for $\xi 1-b$ whenever $b_R(\xi)$ converges in A. The last statement of the lemma follows from the fact that left and right inverses coincide when both exist.

5.5 Lemma. Let A be a Banach algebra with identity over C; let $a \in A$ and $\xi \in \rho(a)$. Then the spectral radius of $R(\xi;a)$ is given by

$$r(R(\xi;a)) = 1/\mathrm{dist}\{\xi;\sigma(a)\}.$$

Proof. Let A_0 be the second commutant of a in A. It is a commutative closed subalgebra of A containing 1 and a; the spectrum of a as an element of A_0 coincides with its spectrum $\sigma(a)$ (relative to A), and $R(\xi;a) \in A_0$ for all $\xi \in \rho(a)$. Let M be the maximal ideal space of the commutative Banach algebra A_0, and let $x \to \hat{x}$ denote the Gelfand transform of A_0 into $C(M)$. Then

$$r(R(\xi;a)) = \sup_{m\in M} |R(\xi;a)^\wedge(m)|$$

$$= \sup_{m\in M} |\xi-\hat{a}(m)|^{-1}$$

$$= \sup_{\lambda\in\sigma(a)} |\xi-\lambda|^{-1} = 1/\inf_{\lambda\in\sigma(a)} |\xi-\lambda|$$

$$= 1/\mathrm{dist}\{\xi;\sigma(a)\}.$$

5.6 Notation. For $a,b \in A$ (a Banach algebra with identity 1), set

$$r(a,b) = \limsup_{n\to\infty} \|C(a,b)^n 1\|^{1/n} \qquad (5.6.1)$$

Clearly $r(a,b) \leq r(C(a,b))$ (the spectral radius of the operator $C(a,b)$ acting in A).

We write

$$\sigma_L(a,b) = \{\xi \in C;\ \mathrm{dist}\{\xi,\sigma(a)\} \leq r(a,b)\}$$

$$\sigma_R(a,b) = \{\xi \in C;\ \mathrm{dist}\{\xi,\sigma(a)\} \leq r(b,a)\}$$

and

$$\sigma(a,b) = \sigma_L(a,b) \cup \sigma_R(a,b)$$
$$= \{\xi \in C;\ \mathrm{dist}\{\xi,\sigma(a)\} \leq d(a,b)\}$$

where

$$d(a,b) = \max\{r(a,b), r(b,a)\}.$$

5.7 Lemma. Let a,b be elements of a Banach algebra A with identity 1 over C. Then the series $b_L(\xi)$ $(b_R(\xi))$ converges absolutely and uniformly on every compact subset of $C \setminus \sigma_L(a,b)$ $(C \setminus \sigma_R(a,b))$ and is a left (right) inverse of $\xi 1 - b$ for each ξ in the corresponding domain. In particular

$$\sigma(b) \subset \sigma(a,b)$$

and $R(\xi;b) = b_L(\xi) = b_R(\xi)$, $\qquad \xi \in C \setminus \sigma(a,b)$.

Proof. If $\xi \in C \setminus \sigma_L(a,b)$, then certainly $\xi \in \rho(a)$ and

$$\limsup_{j\to\infty} \|R(\xi;a)^{j+1} \cdot C(a,b)^j 1\|^{1/j} \leq \lim_{j\to\infty} \|R(\xi;a)^j\|^{1/j} \limsup_{j\to\infty} \|C(a,b)^j 1\|^{1/j}$$
$$= r(R(\xi;a))\, r(a,b)$$
$$= r(a,b)/\mathrm{dist}\{\xi,\sigma(a)\} < 1$$

by Lemma 5.5, so that $b_L(\xi)$ converges absolutely (by the root test), and is therefore a left inverse for $\xi 1 - b$ (by Lemma 5.4). The uniform (absolute) convergence of $b_L(\xi)$ on compact subsets of $C \setminus \sigma_L(a,b)$ is easily verified by direct estimates of $\|R(\xi;a)^{j+1}\|$ (this is left to the reader). The claim about $b_R(\xi)$ is proved in the same way, and the proof is concluded by referring to the last statement of Lemma 5.4.

5.8 Theorem. Let a,b be elements of a Banach algebra A with identity 1 over C. Then for each $f \in \mathcal{H}(\sigma(a,b))$, the following "Taylor formula" is valid

$$f(b) = \sum_{j=0}^{\infty} (-1)^j \frac{f^{(j)}(a)}{j!} C(a,b)^j 1$$

$$= \sum_{j=0}^{\infty} C(b,a)^j 1 \frac{f^{(j)}(a)}{j!}$$

(the series converge absolutely in A).

Proof. Let $f \in H(\sigma(a,b))$, and let $\Gamma = \Gamma(\Omega_f, \sigma(a,b))$ (cf. §1), where Ω_f denotes the neighborhood of $\sigma(a,b)$ in which f is analytic.

By Lemma 5.7, $\sigma(a,b) \supset \sigma(b)$ and therefore

$$f(b) = \frac{1}{2\pi i} \int_\Gamma f(\xi) b_L(\xi) d\xi = \frac{1}{2\pi i} \int_\Gamma f(\xi) b_R(\xi) d\xi.$$

Integrating the series term by term (cf. Lemma 5.7), the result follows.

Note that $f(b)$ and $f^{(j)}(a)$ make sense through the analytic operational calculus, since $\sigma(a) \cup \sigma(b) \subset \sigma(a,b)$.

5.9 Remarks. 1. If a,b are elements of an arbitrary algebra with identity 1, Theorem 5.8 is valid for polynomials f (of course, the series reduce to finite sums in that case). By linearity (with respect to f), it suffices to verify the identities for $f(\xi) = \xi^k$. Now, since L_a commutes with $C(a,b)$, we have

$$b^k = R_b{}^k 1 = [L_a - C(a,b)]^k 1$$

$$= \sum_{j=0}^{k} (-1)^j \binom{k}{j} L_a{}^{k-j} C(a,b)^j 1.$$

Similarly, R_a commutes with $C(b,a)$, hence

$$b^k = L_b{}^k 1 = [R_a + C(b,a)]^k 1$$

$$= \sum_{j=0}^{k} \binom{k}{j} R_a{}^{k-j} C(b,a)^j 1.$$

These are precisely the formulas of Theorem 5.8 for $f(\xi) = \xi^k$.

2. When a,b are commuting elements, since $C(a,b)^j 1 = (a-b)^j$, we have (in the Banach algebra context)

$$r(a,b) = r(a-b) = r(b-a) = r(b,a) = d(a,b)$$

where $r(x)$ denotes as usual the spectral radius of x.

Thus $\sigma(a,b) = \{\xi \in C; \mathrm{dist}\{\xi,\sigma(a)\} \leq r(b-a)\}$,

$$R(\xi,b) = \sum_{j=0}^{\infty} (-1)^j R(\xi;a)^{j+1}(a-b)^j$$

$$= \sum_{j=0}^{\infty} (b-a)^j R(\xi;a)^{j+1} \qquad \text{for} \qquad \xi \in C \setminus \sigma(a,b),$$

and

$$f(b) = \sum_{j=0}^{\infty} \frac{f^{(j)}(a)}{j!}(b-a)^j \qquad \text{for} \qquad f \in H(\sigma(a,b)). \tag{5.9.1}$$

In particular, when $b-a$ is quasi-nilpotent (i.e., $r(b-a) = 0$), we have $\sigma(a,b) = \sigma(a)$, and therefore $\sigma(b) \subset \sigma(a)$ by Lemma 5.7. By symmetry ($r(a-b) = 0$ as well!), we conclude that $\sigma(a) = \sigma(b)$; the resolvent formulas above are valid for all $\xi \in \rho(a) = \rho(b)$, and the Taylor formula (5.9.1) applies to all $f \in H(\sigma(b))$ ($= H(\sigma(a))$).

In general, we clearly have $\sigma(a,b) = \sigma(a)$ if and only if $d(a,b) = 0$. Thus, if $d(a,b) = 0$, one has $\sigma(b) \subset \sigma(a)$ by Lemma 5.7, and so, by symmetry ($d(a,b) = d(b,a)$!), $\sigma(b) = \sigma(a)$. Formally:

5.10 Corollary. Let a,b be elements of a Banach algebra with identity. If $d(a,b) = 0$, then $\sigma(a) = \sigma(b)$.

Actually, $d(\cdot,\cdot)$ is a pseudo-metric, called the "spectral pseudo-metric" [47].

Elements a,b equivalent with respect to the spectral pseudo-metric ($d(a,b) = 0$) are said to be "spectrally equivalent" [47] or "quasi-nilpotent equivalent" [4], and one denotes this (equivalence) relation by $a \sim b$ (see [47] for details).

5.11 Volterra elements.

Throughout this section, A will denote a complex Banach algebra with identity.

Definition. Let $s \in A$. An element $v \in A$ is s-Volterra if

$$[s,v] = v^2 . \tag{5.11.1}$$

It is evident that if v is s-Volterra, then it is also $(s+v')$-Volterra for any $v' \in A$ commuting with v. If $a \in A$ commutes with both s and v, then av is as-Volterra.

More general commutation relations can be reduced to (5.11.1). For example, if $v \in A$ satisfies

$$[s,v] = \varphi(v)$$

with $\varphi \in H(\sigma(v))$ fixed, and the differential equation

$$\varphi f' - f^2 = 0$$

has a solution $f \in H(\sigma(v))$, then $f(v)$ is s-Volterra. Indeed, by Lemma 5.1,

$$[s,f(v)] = D_s f(v) = f'(v) D_s v = (\varphi f')(v) = f^2(v) = [f(v)]^2.$$

This is the case for the commutation relation

$$[s,v] = v^k \qquad (k \geq 2 \text{ fixed}),$$

since the differential equation $z^k f' - f^2 = 0$ has the entire solution $f(z) = (k-1)z^{k-1}$; thus $(k-1)v^{k-1}$ is s-Volterra.

The classical Volterra operator J acting in $C[\alpha,\beta]$ $(0 \in [\alpha,\beta])$ is M-Volterra, where M is the classical multiplication operator. Similarly, in the Hardy space $H^2(D)$ of the unit disc D, the operator $v: f(z) \to \int_0^z f(\xi)d\xi$ (the path of integration from 0 to z is immaterial) is s-Volterra for $s: f(z) \to zf(z)$. With respect to the standard basis $\{z^n;\ n = 0,1,\ldots\}$, the pair (s,v) is a pair of weighted shifts

$$s: z^n \to z^{n+1}$$
$$v: z^n \to \frac{1}{n+1} z^{n+1}, \qquad n = 0,1,2,\ldots$$

More generally, if $A = B(H)$ is the Banach algebra of all bounded linear operators on the Hilbert space H with basis $\{e_n;\ n = 0,1,2,\ldots\}$, and s,v are weighted shifts with non-zero weights $\{\sigma_n\}$ and $\{\omega_n\}$ respectively (that is, $se_n = \sigma_n e_{n+1}$ and $ve_n = \omega_n e_{n+1}$; $n = 0,1,\ldots$), then the Volterra commutation relation (5.11.1) becomes

$$\sigma_{n+1}\omega_n - \omega_{n+1}\sigma_n = \omega_n\omega_{n+1}$$

i.e.,

$$\sigma_{n+1}/\omega_{n+1} = 1 + \sigma_n/\omega_n \qquad n = 0,1,\ldots$$

Set $\lambda = \sigma_0/\omega_0$ arbitrary. The solution of the above recurrence relation is then $\sigma_n/\omega_n = n+\lambda$. Thus, the most general weighted shift v which is s-Volterra (for the given weighted shift s) has the weights

$$\omega_n = \sigma_n/(n+\lambda) \qquad n = 0,1,2,\ldots$$

with λ complex arbitrary $\neq 0,-1,-2,\ldots$. In particular, the same s can possess many s-Volterra associates v.

We collect a few elementary relations:

5.12 Lemma. Let $s,v \in A$, and suppose v is s-Volterra. Then, for $\alpha,\beta \in C$ $n = 0,1,2,\ldots$,

(1) $[s,v^n] = nv^{n+1}$

(2) $v^{n+1} = D_s{}^n v/n!$

(3) $C(s+\alpha v, s+\beta v)^n 1 = (-1)^n n!\binom{\beta-\alpha}{n} v^n$.

Proof. (1) is the special case of (5.1.1) with $D = D_s$. Assuming (2) for $n-1$ ($n \geq 1$), (1) implies $D_s{}^n v = D_s(D_s{}^{n-1}v) = (n-1)!D_s v^n = n!v^{n+1}$, and (2) follows by induction.

(3) is trivial for $n = 0$. Assuming (3) for n, we have by (1)

$$\begin{aligned} C(s+\alpha v,s+\beta v)^{n+1}1 &= (-1)^n n!\binom{\beta-\alpha}{n}\{(s+\alpha v)v^n - v^n(s+\beta v)\} \\ &= (-1)^{n+1}n!\binom{\beta-\alpha}{n}(\beta-\alpha-n)v^{n+1} \\ &= (-1)^{n+1}(n+1)!\binom{\beta-\alpha}{n+1}v^{n+1}, \end{aligned}$$

and (3) follows by induction.

5.13 Lemma. Let v be s-Volterra. Then

(1) v is quasi-nilpotent (in fact, $||v^n||^{1/n} = O(1/n)$).

(2) For $\alpha,\beta \in C$,

$$r(s+\alpha v, s+\beta v) = \begin{cases} 0 & \text{if } \beta-\alpha \text{ is a non-negative integer} \\ \limsup(n!||v^n||)^{1/n} & \text{otherwise.} \end{cases}$$

(3) $d(s+\alpha v,s+\beta v) = \limsup(n!||v^n||)^{1/n}$ for $\alpha \neq \beta$.

(4) $s+\alpha v \sim s+\beta v$ for $\alpha \neq \beta$ if and only if $(n!||v^n||)^{1/n} \to 0$ (that is $||v^n||^{1/n} = o(1/n)$).

(5) $d(s+\alpha v,s+\beta v) \leq \operatorname{diam} \sigma(s)$.

Proof. (1) By Lemma 5.12(1), $||v^n|| \leq 2||s||||v^{n-1}||/(n-1)$ for $n \geq 2$, and therefore by induction

$$||v^n|| \leq ||v||(2||s||)^{n-1}/(n-1)! \qquad (n \geq 2).$$

Hence $||v^n||^{1/n} = O(1/n)$ (By Stirling's formula).

In particular, v is quasi-nilpotent.

(2) If $\beta-\alpha$ is a non-negative integer, $\binom{\beta-\alpha}{n} = 0$ for n large enough; hence $r(s+\alpha v, s+\beta v) = 0$ by Lemma 5.12(3). Otherwise, $|\binom{\beta-\alpha}{n}|^{1/n} \to 1$ as $n \to \infty$ (since $\binom{z}{n} = (-1)^n(n-1)^{-z}/n\Gamma_{n-1}(-z)$, where $\Gamma_n(-z) \to \Gamma(-z)$ when z is not a non-negative integer; cf. [3; p.286]), and (2) follows from Lemma 5.12(3).

(3) and (4) are obvious consequences of (2).

(5) By (2) and Lemma 5.12(2),

$$r(s+\alpha v, s+\beta v) = \begin{cases} 0 \text{ if } \beta-\alpha \text{ is a non-negative integer} \\ \limsup ||D_s{}^n v||^{1/n} \text{ otherwise}. \end{cases}$$

However

$$\limsup ||D_s{}^n v||^{1/n} \leq \limsup ||D_s{}^n||^{1/n} ||v||^{1/n} = r(D_s)$$

where $r(D_s)$ denotes the spectral radius of the operator $D_s \in B(A)$. Since $D_s = L_s - R_s$, where L_s and R_s commute,

$$\sigma(D_s) \subset \sigma(L_s) - \sigma(R_s) \subset \sigma(s) - \sigma(s).$$

Hence

$$r(D_s) \leq \sup\{|\lambda-\mu|;\ \lambda,\mu \in \sigma(s)\} = \operatorname{diam} \sigma(s),$$

and (5) follows.

5.14 <u>Corollary</u>. Let s be quasi-nilpotent and let v be s-Volterra. Then $s+\alpha v \sim s+\beta v$ for all $\alpha, \beta \in \mathbb{C}$.

<u>Proof.</u> Apply Lemma 5.13(5).

An example of this situation is obtained by taking the weighted shifts mentioned above, with $\sigma_n \downarrow 0$.

The condition $||v^n||^{1/n} = o(1/n)$ being independent of α, β, we conclude from Lemma 5.13(4):

5.15 <u>Corollary</u>. Let v be s-Volterra. Then $s+\alpha v \sim s+\beta v$ for either all or no pair $(\alpha, \beta) \in \mathbb{C}^2$, $\alpha \neq \beta$.

By Lemma 5.13(4) and Corollary 5.10, we have

5.16 <u>Corollary</u>. If v is s-Volterra and $||v^n||^{1/n} = o(1/n)$, then $\sigma(s+\alpha v) = \sigma(s)$ for all $\alpha \in \mathbb{C}$.

5.17 <u>Example</u>. Consider M, J acting in $C[0,1]$. A simple calculation shows that $n!||J^n|| = 1$ for all n. Thus, by Lemma 5.13(3),

$$d(M+\alpha J, M+\beta J) = 1 \qquad \text{when } \alpha \neq \beta.$$

In particular, $M+\alpha J \sim M+\beta J$ if and only if $\alpha = \beta$. The same is true when M,J act in $L^p(0,1)$, $1 \leq p \leq \infty$.

By Lemma 5.13(2), $r(s+\alpha v, s+\beta v)$ is symmetric if $\beta-\alpha$ is not an integer. On the other hand, if $\beta-\alpha$ is an integer $\neq 0$ and $\limsup(n!||v^n||)^{1/n} > 0$, then

$$r(s+\alpha v, s+\beta v) \neq r(s+\beta v, s+\alpha v).$$

For the pair (M,J) for example, one side is 0 while the other is 1.

5.18 Corollary. Let v be s-Volterra. Then

$$R(\lambda; v) = \lambda^{-1}1 + \lambda^{-2}e^{s/\lambda}ve^{-s/\lambda} \qquad (\lambda \neq 0).$$

Proof. Since v is quasi-nilpotent (Lemma 5.13(1)), we have for all $\lambda \neq 0$ (by Lemma 5.12(2) and the commutativity of L_s and R_s):

$$R(\lambda; v) = \sum_{n=0}^{\infty} \lambda^{-n-1}v^n = \lambda^{-1}1 + \lambda^{-2}\sum_{n=1}^{\infty} \lambda^{-(n-1)}D_s^{\,n-1}v/(n-1)!$$

$$= \lambda^{-1}1 + \lambda^{-2}\exp(D_s/\lambda)v$$
$$= \lambda^{-1}1 + \lambda^{-2}\exp(L_s/\lambda)\exp(-R_s/\lambda)v$$
$$= \lambda^{-1}1 + \lambda^{-2}e^{s/\lambda}ve^{-s/\lambda}.$$

Applying Lemmas 5.4, 5.7, 5.12(3) and 5.13(3) and Theorem 5.8, we obtain

5.19 Theorem. Let A be a complex Banach algebra with identity 1. Let $s,v \in A$, and suppose v is s-Volterra. For each $\alpha \in C$ and $\zeta \in \rho(s)$, consider the formal series

$$b_L(\zeta;\alpha) = \sum_{j=0}^{\infty} \binom{\alpha}{j} j!R(\zeta;s)^{j+1}v^j$$

$$b_R(\zeta;\alpha) = \sum_{j=0}^{\infty} (-1)^j\binom{-\alpha}{j} j!v^j(\zeta;s)^{j+1}.$$

Then

(1) If $b_L(\zeta;\alpha)$ $(b_R(\zeta;\alpha))$ converges in A, it is a left (right) inverse for $\zeta 1 - (s+\alpha v)$. In particular, if both converge, then $\zeta \in \rho(s+\alpha v)$ and

$$R(\zeta; s+\alpha v) = b_L(\zeta;\alpha) = b_R(\zeta;\alpha). \qquad (5.19.1)$$

(2) Let $\sigma^*(s) = \{\zeta \in C;\ \text{dist}\{\zeta;\sigma(s)\} \leq \limsup(n!||v^{-n}||)^{1/n}\}$. Then $\sigma(s+\alpha v) \subset \sigma^*(s)$ for each $\alpha \in C$, and (5.19.1) is valid for all $\zeta \notin \sigma^*(s)$,

with both series absolutely and uniformly convergent in every compact subset of $C\setminus\sigma^*(s)$.

(3) For each $f \in H(\sigma^*(s))$ and $\alpha \in C$,

$$f(s+\alpha v) = \sum_{j=0}^{\infty} \binom{\alpha}{j} f^{(j)}(s) v^j = \sum_{j=0}^{\infty} (-1)^j \binom{-\alpha}{j} v^j f^{(j)}(s),$$

with both series converging absolutely.

5.20 Corollary. (the exponential formula for Volterra elements). Let v be s-Volterra Then for all $\lambda,\alpha \in C$,

$$e^{\lambda(s+\alpha v)} = e^{\lambda s}(1+\lambda v)^{\alpha} = (1-\lambda v)^{-\alpha} e^{\lambda s}.$$

Proof. Take $f(\xi) = e^{\lambda\xi}$ in Theorem 5.19(3) :

$$e^{\lambda(s+\alpha v)} = e^{\lambda s} \sum_{j=0}^{\infty} \binom{\alpha}{j} (\lambda v)^j = e^{\lambda s}(1+\lambda v)^{\alpha}$$

$$= \sum_{j=0}^{\infty} \binom{-\alpha}{j} (-\lambda v)^j e^{\lambda s} = (1-\lambda v)^{-\alpha} e^{\lambda s}.$$

Note that $(1+\lambda v)^{\alpha}$ makes sense for all $\lambda,\alpha \in C$ since v is quasi-nilpotent (cf. Lemma 5.13(1)).

Remark. For any $a,b \in A$, since L_a commutes with $C(a,b)$ (and, similarly, since R_a commutes with $C(b,a)$), we have

$$e^{\lambda b} = \exp(\lambda R_b)1 = \exp[\lambda L_a - \lambda C(a,b)]1$$

$$= \exp\lambda L_a \exp[-\lambda C(a,b)]1 \tag{5.20.1}$$

and similarly

$$e^{\lambda b} = \exp(\lambda L_b)1 = \exp[\lambda R_a + \lambda C(b,a)]1$$

$$= \exp\lambda R_a \exp[\lambda C(b,a)]1 . \tag{5.20.2}$$

For $a = s$ and $b = s+\alpha v$ with v s-Volterra, Lemma 5.12(3) implies that

$$e^{\lambda(s+\alpha v)} = e^{\lambda s} \sum_{n=0}^{\infty} (-1)^n \lambda^n C(s,s+\alpha v)^n 1/n!$$

$$= e^{\lambda s} \sum_{n=0}^{\infty} \binom{\alpha}{n} \lambda^n v^n = e^{\lambda s}(1+\lambda v)^{\alpha}.$$

The second exponential formula of Corollary 5.20 follows in a similar fashion from (5.20.2).

These exponential formulas will be an important tool in the spectral analysis of the family $\{s+\alpha v;\ \alpha \in C\}$.

5.21 Corollary. Let v be s-Volterra, and suppose $\rho(s)$ is connected. Then for all integers k,

(a) $\sigma(s+kv) \subset \sigma(s)$;

(b) for all $\zeta \in \rho(s)$,

$$R(\zeta;s+kv) = \sum_{j=0}^{k} \binom{k}{j} j!R(\zeta;s)^{j+1}v^j, \quad k \geq 0$$

$$= \sum_{j=0}^{|k|} (-1)^j \binom{|k|}{j} j!v^j R(\zeta;s)^{j+1}, \quad k < 0;$$

(c) for each $f \in H(\sigma(s))$,

$$f(s+kv) = \sum_{j=0}^{k} \binom{k}{j} f^{(j)}(s)v^j, \quad k \geq 0$$

$$= \sum_{j=0}^{|k|} (-1)^j \binom{|k|}{j} v^j f^{(j)}(s), \quad k < 0.$$

Proof. By Theorem 5.19(2), we have

(*) $b_L(\xi;k)[\xi 1-(s+kv)] = [\xi 1-(s+kv)]b_L(\xi;k) = 1$ for all $\xi \in C \smallsetminus \sigma^*(s) \subset \rho(s)$.

If k is a non-negative integer, $b_L(\xi;k)$ reduces to a finite sum, and therefore all sides in (*) are holomorphic in $\rho(s)$. Since $\rho(s)$ is assumed connected, (*) is valid for all $\xi \in \rho(s)$. This proves (a) and (b) for $k \geq 0$, and (c) follows from (b) and the Riesz-Dunford integral. When k is a negative integer, we apply the above argument to $b_R(\xi;k)$ instead of $b_L(\xi;k)$.

5.22 Corollary. Let k be an integer such that both $\rho(s)$ and $\rho(s+kv)$ are connected. Then $\sigma(s+kv) = \sigma(s)$.

Proof. Since $\rho(s)$ is connected, we have $\sigma(s+kv) \subset \sigma(s)$ by Corollary 5.21(a). Set $s' = s+kv$. Since v is s'-Volterra and $\rho(s')$ is connected, we may apply Corollary 5.21(a) with $(s',v,-k)$ replacing (s,v,k); whence $\sigma(s) = \sigma(s'-kv) \subset \sigma(s') = \sigma(s+kv)$.

5.23 Corollary. Suppose $\sigma(s)$ lies on a finite union of Jordan arcs Γ that does not separate the plane. Let v be s-Volterra. Then $\sigma(s+kv) = \sigma(s)$ for all $k \in Z$.

Proof. By Corollary 5.21(a), $\sigma(s+kv)$ lies also on Γ, hence both $\rho(s)$ and $\rho(s+kv)$ are connected.

The special case of interest in the sequel is when $\sigma(s) \subset \mathbb{R}$.

5.24 Corollary. Let s be of class C^n with real spectrum, and let v be s-Volterra. Then for each $k \in \mathbb{Z}$, $s+kv$ is of class $C^{n+|k|}$. Moreover, if τ_s denotes the $C^n[\alpha,\beta]$-o.c. for s, then the $C^{n+|k|}[\alpha,\beta]$-o.c. for $s+kv$ is given by

$$\tau_{s+kv}(f) = \sum_{j=0}^{k} \binom{k}{j} \tau_s(f^{(j)}) v^j, \qquad k \geq 0$$

$$= \sum_{j=0}^{|k|} (-1)^j \binom{|k|}{j} v^j \tau_s(f^{(j)}) \qquad k < 0$$

$(f \in C^{n+|k|}[\alpha,\beta])$. (5.24.1)

Proof. The map τ_{s+kv} defined above is continuous and linear on $C^{n+|k|}[\alpha,\beta]$ into the given Banach algebra A. By Corollary 5.21(c), it coincides with the analytic operational calculus when restricted to polynomials. In particular, it is multiplicative on polynomials, hence on $C^{n+|k|}[\alpha,\beta]$, by density.

5.25. Examples. 1. Let $[\alpha,\beta]$ be any closed interval containing 0, and let X be either $L^p[\alpha,\beta]$ $(1 \leq p \leq \infty)$ or $C[\alpha,\beta]$. In $A = B(X)$, take $s = M$ and $V = J$. Since M is of class C with $\sigma(M) = [\alpha,\beta]$ and $C[\alpha,\beta]$-o.c. $\tau_0(f): \varphi(x) \to f(x)\varphi(x)$ $(f \in C[\alpha,\beta], \varphi \in X)$, Corollaries 5.23 and 5.24 imply that $M+kJ$ is of class $C^{|k|}$ with $\sigma(M+kJ) = [\alpha,\beta]$, and its $C^{|k|}[\alpha,\beta]$-o.c. is given by

$$\tau_k(f) = \sum_{j=0}^{k} \binom{k}{j} \tau_0(f^{(j)}) J^j = J^{-k}\tau_0(f) J^k \qquad (k \geq 0)$$

$$= \sum_{j=0}^{|k|} (-1)^j \binom{|k|}{j} J^j \tau_0(f^{(j)}) \qquad (k < 0).$$

The case $k \geq 0$ was discussed directly in §2.2. A similar approach is possible for $k < 0$ (cf. [23], Lemmas 4 and 5).

2. Let $A = B(X)$ with $X = C[\alpha,\beta]$. Let σ be a real continuous function of bounded variation in $[\alpha,\beta]$. Define $M_\sigma: \varphi(x) \to \sigma(x)\varphi(x)$ and $J_\sigma: \varphi(x) \to \int_\alpha^x \varphi(t)\,d\sigma(t)$ $(\varphi \in C[\alpha,\beta])$. We have

$$(J_\sigma^2\varphi)(x) = \int_\alpha^x \int_\alpha^t \varphi(s)\,d\sigma(s)\,d\sigma(t) = \int_\alpha^x \left(\int_s^x d\sigma(t)\right)\varphi(s)\,d\sigma(s)$$

$$= \int_\alpha^x [\sigma(x)-\sigma(s)]\varphi(s)\,d\sigma(s) = ([M_\sigma, J_\sigma]\varphi)(x)$$

for all $x \in [\alpha,\beta]$ and $\varphi \in C[\alpha,\beta]$. Thus J_σ is M_σ-Volterra. Since M_σ is clearly of class C, Corolalry 5.24 implies that $M_\sigma+kJ_\sigma$ is of class $C^{|k|}$ $(k \in \mathbb{Z})$, and its $C^{|k|}[\alpha,\beta]$-o.c. is given by (5.24.1).

6. The family $S+\zeta V$

The role played by the concrete operators $T_n = M+nJ$ $(n = 0,1,2,\ldots)$ in the general theory of C^n-operators motivates the following problem. Given a non-negative integer n, determine the set of $\zeta \in C$ for which the operator $T_\zeta = M+\zeta J$ (acting in $L^p(\Delta)$ or $C(\Delta)$) is of class C^n.

We shall solve this problem and some related ones in the abstract context of the preceding section, with $A = B(X)$ for a fixed Banach space X.

Some tools from the theory of semigroups of operators will be needed. They will be described briefly, and the reader is referred to [11] for proofs and details.

Let C^+ denote the right half-plane $\{\zeta \in C; \operatorname{Re}\zeta > 0\}$, and let $V(\cdot): C^+ \to B(X)$ be holomorphic and satisfy the semigroup identity $V(\zeta+\omega) = V(\zeta)V(\omega)$ for all $\zeta,\omega \in C^+$. We then say that $V(\cdot)$ is a holomorphic semigroup in C^+. It is of class C_0 if $V(\xi) \to I$ strongly as $R \ni \xi \to 0+$.

6.0 Definition. A regular semigroup is a holomorphic semigroup of class C_0 in C^+, which is bounded on the rectangle $Q = \{\zeta = \xi+i\eta;\ 0 < \xi \le 1,\ |\eta| \le 1\}$.

We set

$$\nu = \nu(V) = \sup_Q \log||V(\zeta)|| .$$

Thus $0 \le \nu < \infty$.

6.1 Theorem. Let $V(\cdot)$ be a regular semigroup. Then for each $\eta \in R$, $V(\xi+i\eta)$ converges strongly as $\xi \to 0+$ to a bounded operator $V(i\eta)$ with the following properties:

(1) $\{V(i\eta);\ \eta \in R\}$ is a strongly continuous group;

(2) $V(i\eta)$ commutes with $V(\zeta)$ for all $\eta \in R, \zeta \in C^+$;

(3) $V(\xi+i\eta) = V(\xi)V(i\eta)$ $(\xi > 0, \eta \in R)$;

(4) $V(\cdot)$ is of exponential type $\le \nu$ in $\overline{C^+}$, that is, there exists a constant $K > 0$ such that

$$||V(\zeta)|| \le Ke^{\nu|\zeta|} \qquad (\zeta \in \overline{C^+}).$$

Proof. See [11], Theorem 17.9.1 and its proof.

6.2 Theorem. Let $V(\cdot)$ be a regular semigroup. Then for each $\zeta \in C^+$, $V(\zeta)$ is one-to-one with dense range.

Proof. Since $\{V(i\eta);\ \eta \in R\}$ is a group of operators, the theorem needs proof only for $\zeta \in C^+$.

Fix a positive integer k such that $\nu/k < \pi$. Suppose $V(1/k)x = 0$ for some $x \in X$. Set $x(\zeta) = V(\zeta/k)x$ in $\overline{C^+}$. Then $x(\cdot)$ is strongly continuous in $\overline{C^+}$, and is holomorphic of exponential type $< \pi$ in C^+. Moreover, $x(n) = V(1/k)^n x = 0$ for $n = 1,2,\ldots$. By Theorem 3.13.7 in [11], $x(\zeta) = 0$ for all $\zeta \in \overline{C^+}$; in particular, $x = x(0) = 0$. This shows that $V(1/k)$ is one-to-one.

Next, for $\zeta \in C^+$ arbitrary, suppose $V(\zeta)x = 0$ for some $x \in X$. By Theorem 6.1, $V(\xi)x = 0$, where $\xi = \mathrm{Re}\zeta$. Let n be any integer $\geq \xi$. Then

$$V(1/k)^{nk}x = V(n)x = V(n-\xi)V(\xi)x = 0.$$

Since $V(1/k)$ is one-to-one, it follows that $x = 0$, that is, $V(\zeta)$ is one-to-one.

The conjugate semigroup $\{V(\zeta)^*;\ \zeta \in C^+\}$ is holomorphic with $\nu(V^*) = \nu < \infty$; it is of class C_0 on the subspace

$$X_0^* = \text{closure}\ \cup\{V(\xi)^*X^*;\ \xi > 0\}.$$

Therefore $V(\zeta)^*$ is one-to-one on X_0^*, for each given $\zeta \in C^+$. Suppose $V(\xi+i\eta)^*x^* = 0$ for some $x^* \in X^*$, $\xi > 0$ and $\eta \in R$. Then $V(i\eta)^*V(\xi)^*x^* = 0$, and so $V(\xi)^*x^* = 0$ since $V(i\eta)$, and hence $V(i\eta)^*$, is nonsingular. For each $n = 2,3,\ldots$, $V(\xi - \xi/n)^*V(\xi/n)^*x^* = 0$, and since $V(\xi/n)^*x^* \in X_0^*$ and $V(\xi-\xi/n)^*$ is one-to-one on X_0^*, it follows that $V(\xi/n)^*x^* = 0$. Hence $x^*V(\xi/n)x = 0$ for each $x \in X$ and $n = 2,3,\ldots$, and therefore $x^*x = 0$ for each $x \in X$ (because $V(\cdot)$ is of class C_0). This shows that $x^* = 0$; hence $V(\xi+i\eta)^*$ is one-to-one, and so $V(\xi+i\eta)$ has dense range.

6.3 For $\mathrm{Re}\zeta < 0$ define the unbounded operator $V(\zeta)$ as $V(-\zeta)^{-1}$. By Theorems 6.2 and 6.1, $V(\zeta)$ is a well-defined closed operator with dense domain $\mathcal{D}V(\zeta) = V(-\zeta)X = V(-\xi)X$ (where $\zeta = \xi+i\eta$). One verifies easily that

$$V(\zeta+\omega) = V(\zeta)V(\omega)$$

for all $\zeta,\omega \in C^- = \{\zeta \in C; \mathrm{Re}\zeta < 0\}$; $\mathcal{D}V(\zeta) \subset \mathcal{D}V(\omega)$ for $\mathrm{Re}\zeta \leq \mathrm{Re}\omega < 0$; and if $x \in X_0 = \cup\{\mathcal{D}V(\zeta), \mathrm{Re}\zeta < 0\}$, then $V(\xi+i\eta)x \to V(i\eta)x$ strongly as $\xi \to 0-$ and $V(\cdot)x$ is holomorphic in a half-plane (specifically, the half-plane $\mathrm{Re}\zeta > -\alpha$ when $x \in \mathcal{D}V(-\alpha)$, $\alpha > 0$). Thus, a regular semigroup $V(\cdot)$ extends as a function from C to closed densely defined operators, such that $\{V(\zeta);\zeta \in C^-\}$ is itself a "holomorphic semigroup" with $\{V(i\eta);\eta \in R\}$ as its boundary group, in the sense described above. The notation $V(\cdot)$ will be used for this extended function.

By Theorem 6.1(4), the boundary group $\{V(i\eta);\eta \in R\}$ satisfies the growth condition

$$||V(i\eta)|| \leq Ke^{\nu|\eta|} \quad (\eta \in R). \tag{6.3.1}$$

Since $\{V(\xi);\xi \geq 0\}$ itself satisfies a growth condition (cf. [11; pp.306 and 244])

$$||V(\xi)|| \leq K'e^{\omega\xi} \quad (\xi \geq 0), \tag{6.3.2}$$

it follows from (6.3.1) and Theorem 6.1(3) that

$$||V(\zeta)|| \leq K''e^{r\lambda(\theta)} \quad (\zeta = re^{i\theta}) \tag{6.3.3}$$

where $\lambda(\theta) = \omega\cos\theta + \nu|\sin\theta|$ $(-\frac{\pi}{2} \leq \theta \leq \frac{\pi}{2})$. Clearly $\lambda(\theta)$ is bounded, even and

$$\begin{aligned}\limsup_{\delta\to 0+} \delta^{-1}\{\pi-\lambda(\tfrac{\pi}{2}-\delta)\} &= \limsup_{\delta\to 0+}[\delta^{-1}(\pi-\nu\cos\delta) - \omega\sin\delta/\delta] \\ &= \lim_{\delta\to 0+}(\pi-\nu)/\delta = +\infty\end{aligned} \tag{6.3.4}$$

<u>if and only if $\nu < \pi$.</u>

This simple observation may be used for example to prove the following

<u>Proposition</u>. Let $V(\cdot)$ be a regular semigroup whose boundary group satisisies (6.3.1) with $\nu < \pi$. Suppose $A \in B(X)$ commutes with $V = V(1)$. Then A commutes with $V(\zeta)$ for all $\zeta \in R$.

<u>Proof</u>. Set

$$x(\zeta) = AV(\zeta)x-V(\zeta)Ax \quad (\zeta \in \overline{C^+},\ x \in X).$$

For x fixed, $x(\cdot)$ is strongly continuous in $\overline{C^+}$, holomorphic in C^+,

$$||x(i\eta)|| \leq 2K||A||\ ||x||e^{\nu|\eta|} \quad (\eta \in R),$$

$$||x(\zeta)|| \leq 2K''||A||\ ||x||e^{r\lambda(\theta)} \quad (\zeta = re^{i\theta},\ -\tfrac{\pi}{2} \leq \theta \leq \tfrac{\pi}{2})$$

with $\lambda(\theta)$ satisfying the F. Carlson Condition (6.3.4) (since $\nu < \pi$), and finally $x(n) = 0$ for $n = 1,2,\ldots$ (since A commutes with V). By Theorem 3.13.7 in [11], it follows that $x(\zeta) = 0$ for all $\zeta \in \overline{C^+}$, and the proposition follows for $\mathrm{Re}\zeta \geq 0$.

If $\zeta \in C^-$ and $x \in \mathcal{D}V(\zeta)$, then $x = V(-\zeta)y$ and $V(\zeta)x = y$. Hence $Ax = V(-\zeta)Ay \in \mathcal{D}V(\zeta)$ and $V(\zeta)Ax = Ay = AV(\zeta)x$, that is $AV(\zeta) \subset V(\zeta)A$.

Let $S \in B(X)$. We make the following

Standing Hypothesis. (1) $V(\cdot)$ is a regular semigroup whose boundary group satisfies a growth condition

$$||V(i\eta)|| \leq Ke^{\nu|\eta|} \quad (\eta \in R) \tag{6.3.5}$$

with $\nu < \pi$;

(2) $V = V(1)$ is S-Volterra, i.e. $[S,V] = V^2$.

We set

$$T_\zeta = S + \zeta V \qquad (\zeta \in C).$$

6.4 Theorem. For each $\zeta,\alpha \in C$, $\mathcal{D}_\zeta = \mathcal{D}V(\zeta)$ is invariant under S and T_α, and the following equivalent relations are valid on $\mathcal{D}_\zeta$.

(a) $[S,V(\zeta)] = \zeta V(\zeta+1)$;

(b) $SV(\zeta) = V(\zeta)T_\zeta$;

(c) $V(\zeta)S = T_{-\zeta}V(\zeta)$.

Proof. For $x \in X$, set

$$x(\zeta) = [S,V(\zeta)]x - \zeta V(\zeta+1)x , \qquad \zeta \in \overline{C^+}.$$

Clearly, $x(\cdot)$ is strongly continuous in $\overline{C^+}$, holomorphic in C^+, and for $\varepsilon > 0$ fixed such that $\nu+\varepsilon < \pi$, it follows from (6.3.3) that

$$||x(\zeta)|| \leq Ce^{r\lambda'(\theta)} \qquad (\zeta = re^{i\theta}, \ -\tfrac{\pi}{2} \leq \theta \leq \tfrac{\pi}{2})$$

where $\lambda'(\theta) = \lambda(\theta)+\varepsilon$ and C is a suitable constant. By (6.3.4),

$$\limsup_{\delta\to 0+} \delta^{-1}\{\pi-\lambda'(\tfrac{\pi}{2}-\delta)\} = \lim_{\delta\to 0+} (\pi-\nu-\varepsilon)/\delta = +\infty.$$

By Lemma 5.12(1), for $n = 1,2$,

$$x(n) = [S,V^n]x - nV^{n+1}x = 0 .$$

Therefore, by Theorem 3.13.7 in [11], $x(\zeta) = 0$ for all $\zeta \in \overline{C^+}$. Since x was arbitrary, this proves (a) (and hence the trivially equivalent relations (b) and (c)) for $\operatorname{Re}\zeta \geq 0$. In particular, by (b), $S\mathcal{D}_{-\zeta} = SV(\zeta)X = V(\zeta)T_\zeta X \subset V(\zeta)X = \mathcal{D}_{-\zeta}$, i.e., $\mathcal{D}_{-\zeta}$ is invariant for S, and hence also for T_α $(\alpha \in C)$. Moreover, if $x \in \mathcal{D}_{-\zeta}$, say $x = V(\zeta)y$, with $\zeta \in C^+$, then $y = V(-\zeta)x$, and by (b)

$$V(-\zeta)Sx = V(-\zeta)SV(\zeta)y = V(-\zeta)V(\zeta)T_\zeta y = T_\zeta V(-\zeta)x.$$

This proves (c) (and hence all three identities) for $\operatorname{Re}\zeta < 0$.

6.5 Corollary.

For all $\zeta,\omega \in C$,

$$T_\zeta = V(\omega-\zeta)T_\omega V(\zeta-\omega) \quad \text{on } \mathcal{D}_{\zeta-\omega}.$$

In particular, T_ζ is <u>similar</u> to T_ω if $\mathrm{Re}\,\zeta = \mathrm{Re}\,\omega$:

$$T_\zeta = V(i\eta)T_\omega V(-i\eta)$$

with $\eta = \mathrm{Im}(\omega-\zeta)$.

Proof. By Theorem 6.4, we have on $\mathcal{D}_{\zeta-\omega}$

$$T_\omega V(\zeta-\omega) = (T_{\omega-\zeta} + \zeta V)V(\zeta-\omega) = V(\zeta-\omega)(S + \zeta V) = V(\zeta-\omega)T_\zeta,$$

and the corollary follows.

6.6 Remark. For each $\zeta \in C^+$, $V(\zeta)$ is quasi-nilpotent. Indeed, fix a positive integer n such that $\mathrm{Re}\, n\zeta > 1$. Then, by Lemma 5.13(1), $V(\zeta)^n = V(n\zeta-1)V$ is quasi-nilpotent, and therefore $V(\zeta)$ is quasi-nilpotent.

Thus, for each $\eta \in R$, the non-singular operator $V(i\eta)$ is the strong operator limit of the quasi-nilpotent operators $V(\xi+i\eta)$ as $\xi \to 0+$.

6.7 Remark. If $\mathrm{Re}\,\omega \le \mathrm{Re}\,\zeta$, then $\sigma_p(T_\zeta) \subset \sigma_p(T_\omega)$, where $\sigma_p(\cdot)$ denotes the point spectrum. Indeed, let $\lambda \in \sigma_p(T_\zeta)$ and let $x \neq 0$ be a corresponding eigenvector. By Theorem 6.2, $V(\zeta-\omega)x \neq 0$. By Corollary 6.5,

$$T_\omega V(\zeta-\omega)x = V(\zeta-\omega)T_\zeta x = \lambda V(\zeta-\omega)x,$$

hence $\lambda \in \sigma_p(T_\omega)$.

We shall need some norm estimates for the groups $\{\exp(\lambda T_\zeta);\lambda \in C\}$.

6.8 Theorem. There exists a constant $H > 0$ such that

$$||\exp(\lambda T_{\xi+i\eta})|| \le H||e^{\lambda S}||(1+|\lambda|\ ||V||)^{|\xi|}e^{2\nu|\eta|}$$

for all $\xi,\eta \in R$ and $\lambda \in C$.

Proof. By Corollary 6.5,

$$T_{\xi+i\eta} = V(-i\eta)T_\xi V(i\eta).$$

Therefore, by Corollary 5.20

$$\begin{aligned}\exp(\lambda T_{\xi+i\eta}) &= V(-i\eta)\exp(\lambda T_\xi)V(i\eta)\\ &= V(-i\eta)e^{\lambda S}(I+\lambda V)^{\xi}V(i\eta) &(6.8.1)\\ &= V(-i\eta)(I-\lambda V)^{-\xi}e^{\lambda S}V(i\eta)\ . &(6.8.2)\end{aligned}$$

In particular, by (6.3.5)

$$K^{-2}e^{-2\nu|\eta|}||\exp(\lambda T_\xi)|| \le ||\exp(\lambda T_{\xi+i\eta})|| \le K^2e^{2\nu|\eta|}||\exp(\lambda T_\xi)|| \ . \qquad (6.8.3)$$

Fix $\lambda \in C$, and set

$$\Phi_\lambda(\zeta) = e^{\nu\zeta^2}\exp(\lambda T_\zeta) \qquad (\zeta \in C). \qquad (6.8.4)$$

Then

$$||\Phi_\lambda(\xi+i\eta)|| \le K^2||\exp(\lambda T_\xi)||\exp \nu(\xi^2-\eta^2+2|\eta|) \le K^2||\exp(\lambda T_\xi)||\exp \nu(\xi^2+1) . \quad (6.8.4')$$

In particular, for each integer k, $||\Phi_\lambda(\xi+i\eta)||$ is bounded in the strip $k-1 \le \xi \le k$.

By (6.8.1) and (6.8.2),

$$||\Phi_\lambda(k+i\eta)|| \le K^2||e^{\lambda S}||\exp \nu(k^2+1)(1+|\lambda|\ ||V||)^{|k|}$$

for all $\eta \in R$ and $k = 0,\pm1,\pm2,\ldots$. Moreover, the operator-valued function $\Phi_\lambda(\cdot)$ is holomorphic in C. By the "three-lines theorem" (cf. Theorem VI.10.3 in [5]) applied to the operator-valued function $\Phi_\lambda(\cdot)$ in the strip $k-1 \le \xi \le k$ (writing $\xi = t(k-1) + (1-t)k = k-t$ with $0 \le t \le 1$, so that $|\xi| = t|k-1|+(1-t)|k|$), we obtain

$$||\Phi_\lambda(\xi+i\eta)|| \le K^2||e^{\lambda S}||(1+|\lambda|\ ||V||)^{|\xi|}\exp \nu[t(k-1)^2+(1-t)k^2+1].$$

However

$$t(k-1)^2+(1-t)k^2 = (k-t)^2+t(1-t) = \xi^2+t(1-t) \le \xi^2+1/4.$$

Hence

$$||\exp(\lambda T_\xi)|| = e^{-\nu\xi^2}||\Phi_\lambda(\xi)|| \le K^2\ ||e^{\lambda S}||(1+|\lambda|\ ||V||)^{|\xi|}e^{5\nu/4},$$

and the theorem follows from (6.8.3) with $H = K^4e^{5\nu/4}$.

A lower norm estimate is obtained for the groups $\{\exp(it\ T_\zeta);\ t \in R\}$ $(\zeta \in C)$ in case $||e^{itS}|| = O(1)$.

6.9 <u>Theorem.</u> Suppose $||e^{itS}|| = O(1)$ $(t \in R)$. Then there exists a strictly positive (upper semicontinuous) function $C(\cdot)$ on R such that

$$||\exp(it\ T_{\xi+i\eta})|| \ge C(\xi)(1+|t|\ ||V||)^{|\xi|}e^{-2\nu|\eta|}$$

for all $t,\xi,\eta \in R$.

<u>Proof</u>. By (6.8.3),

$$||\exp(it\ T_{\xi+i\eta})|| \ge K^{-2}e^{-2\nu|\eta|}||\exp(it\ T_\xi)||$$

for all $t,\xi,\eta \in R$. (6.8.5)

Let

$$B_t(\xi) = (1+|t|\ ||V||)^{-|\xi|}||\exp(it\ T_\xi)||$$

and

$$B(\xi) = \inf_{t\in R} B_t(\xi), \qquad \xi \in R.$$

Then $B(\cdot)$ is a non-negative upper semicontinuous function, and the theorem will follow from (6.8.5) with $C(\xi) = K^{-2}B(\xi)$ if we show that $B(\xi) \neq 0$ for all ξ.

First, observe that $\sigma(S)$ is real (since $||e^{itS}|| = O(1)$; cf. (3.1.4)), and therefore $B_t(0) = ||e^{itS}|| \geq r(e^{itS}) = 1$, while $B_0(0) = |||I||| = 1$. Hence $B(0) = 1$.

By Corollary 5.20,

$$\exp(it\,T_\xi) = e^{itS}(I+itV)^\xi = (I-itV)^{-\xi}e^{itS}\,. \tag{6.8.6}$$

In particular, for all $t,\xi \in R$,

$$\exp(it\,T_{\xi+1}) = \exp(it\,T_\xi)(I+itV)$$

and

$$\exp(it\,T_{\xi-1}) = (I-itV)\exp(it\,T_\xi).$$

It follows that for all $t \in R$,

$$B_t(\xi+1) \leq B_t(\xi) \qquad (\xi \geq 0)$$

and $B_t(\xi-1) \leq B_t(\xi) \qquad (\xi \leq 0)$.

Hence

$$B(\xi+1) \leq B(\xi) \qquad (\xi \geq 0)$$

and $B(\xi-1) \leq B(\xi) \qquad (\xi \leq 0)$.

Since $B(0) = 1 \neq 0$, the wanted conclusion $(B(\xi) \neq 0$ for all $\xi \in R)$ will now follow if we show that $B(\xi) \neq 0$ for $|\xi| > 1$.

Suppose $B(\xi_0) = 0$ for some real ξ_0 with $|\xi_0| > 1$. Let then $\{t_k\} \subset R$ be such that $B_{t_k}(\xi_0) \to 0$. If $\{t_k\}$ were bounded, then

$$||\exp(it_kT_{\xi_0})|| = (1+|t_k|\,||V||)^{|\xi_0|}B_{t_k}(\xi_0) \to 0 \quad \text{as } k \to \infty.$$

However, if $\{t_k'\}$ is a subsequence of $\{t_k\}$ converging to some $t_0 \in R$, then $||\exp(it_k'T_{\xi_0})|| \to ||\exp(it_0T_{\xi_0})||$. Therefore $\exp(it_0T_{\xi_0}) = 0$, which is impossible since $\exp(it_0T_{\xi_0})$ is non-singular. Hence $\{t_k\}$ is unbounded, and we may assume that $|t_k| \to \infty$.

Given $\varepsilon > 0$, fix k_0 such that

$$B_{t_k}(\xi_0) < \varepsilon^{|\xi_0|} \qquad \text{for } k \geq k_0\,. \tag{6.8.7}$$

For $k \geq k_0$ fixed, consider the entire functions

$$F_k^{\pm}(\zeta) = (1+|t_k|\,||V||)^{\mp\zeta}\,\Phi_{it_k}(\zeta)$$

(cf. (6.8.4)).

They are bounded in each vertical strip $a \le \mathrm{Re}\zeta \le b$, and

$$||F_k^{\pm}(i\eta)|| \le K^2 M e^{\nu} \qquad (\eta \in R) \qquad (6.8.8)$$

where $M = \sup_{t \in R} ||e^{itS}||$.

We consider F_k^+ if $\xi_0 > 0$ and F_k^- if $\xi_0 < 0$. By (6.8.4') and (6.8.7),

$$||F_k^{\pm}(\xi_0+i\eta)|| \le K^2 \exp[\nu(\xi_0^2+1)] B_{t_k}(\xi_0) < K^2 \exp[\nu(\xi_0^2+1)]\varepsilon^{|\xi_0|}. \qquad (6.8.9)$$

We apply the "three-lines theorem" (cf. [5], Theorem VI.10.3) to F_k^+ (resp. F_k^-) in the strip $0 \le \mathrm{Re}\zeta \le \xi_0$ when $\xi_0 > 0$ (resp. $\xi_0 \le \mathrm{Re}\zeta \le 0$ when $\xi_0 < 0$). By (6.8.8) and (6.8.9),

$$||F_k^{\pm}(\xi+i\eta)|| \le K^2 \exp[\nu(\xi_0^2+1)]\varepsilon^{|\xi|}$$

in the respective strips.

Since $|\xi_0| > 1$, this is true in particular for $\xi+i\eta = 1$ (resp. -1). It follows that

$$(1+|t_k|\ ||V||)^{-1}||\exp(it_k T_{\pm 1})|| \le K^2 \exp(\nu\xi_0^2)\varepsilon$$

for all $k \ge k_0$, that is

$$\lim_{k\to\infty}(1+|t_k|\ ||V||)^{-1}||\exp(it_k T_{\pm 1})|| = 0.$$

Since $||e^{itS}|| = O(1)$, we now obtain from (6.8.6) that $\lim_{k\to\infty}(1+|t_k|\ ||V||)^{-1}||I \pm it_k V|| = 0$. However, since $|t_k| \to \infty$, this limit is equal to 1, and we reached the wanted contradiction.

We may solve now the C^k-classification problem for the family of operators $\{T_\zeta; \zeta \in C\}$, in the abstract context of the Standing Hypothesis 6.3.

6.10 Theorem. Let n,k be non-negative integers. Suppose S is of class C^n with real spectrum. Then, for all $\zeta \in C$, $\sigma(T_\zeta) = \sigma(S)$, and T_ζ is of class C^{n+k} for all ζ in the strip $|\mathrm{Re}\zeta| \le k$.

Proof. If $k = 0$ (so that the strip reduces to the imaginary axis), T_ζ is similar to S (by Corollary 6.5) for ζ in the "strip", and the conclusions of the theorem are trivial. Let then k be a positive integer. By Corollaries 5.23 and 5.24, $\sigma(T_{\pm k}) = \sigma(S)$ and $T_{\pm k}$ are of class C^{n+k}. Thus $T_{\pm k} \in [C^{n+k}[\alpha,\beta]]$

for some closed interval $\Delta = [\alpha,\beta]$ containing 0.

Using the notation of Corollary 4.2 and fixing $p \in P$ with $||p||_\Delta \leq 1$, we have

$$||(J^{n+k}p)(T_{\pm k})|| \leq ||T_{\pm k}||_{n+k,\Delta} < \infty \quad . \tag{6.8.10}$$

Set

$$P(\zeta) = \exp(\nu\zeta^2)(J^{n+k}p)(T_\zeta), \qquad \zeta \in C.$$

Clearly, $P(\cdot)$ is entire, and by Corollary 6.5,

$$P(\zeta) = \exp(\nu\zeta^2)V(-i\eta)(J^{n+k}p)(T_\xi)V(i\eta),$$

where $\zeta = \xi+i\eta$.

Hence, by (6.3.5),

$$||P(\zeta)|| \leq K^2\exp[\nu(\xi^2-\eta^2+2|\eta|)]||(J^{n+k}p)(T_\xi)|| \leq K^2\exp[\nu(\xi^2+1)]||(J^{n+k}p)(T_\xi)||$$

Thus $P(\cdot)$ is bounded in the strip $|\xi| \leq k$. Moreover, by (6.8.10),

$$||P(\pm k+i\eta)|| \leq K^2\exp[\nu(k^2+1)]||T_{\pm k}||_{n+k,\Delta}.$$

By the "three lines theorem", it follows that

$$||P(\zeta)|| \leq K^2\exp\ [\nu(k^2+1)] \max\{||T_{-k}||_{n+k,\Delta}, ||T_k||_{n+k,\Delta}\} = K_1$$

for all ζ in the strip $|\mathrm{Re}\zeta| \leq k$. Hence $||(J^{n+k}p)(T_\zeta)|| \leq K_1\exp[\nu(\eta^2-\xi^2)]$, and since $p \in P$ was arbitrary (with $||p||_\Delta \leq 1$), we conclude that

$$||T_{\xi+i\eta}||_{n+k,\Delta} \leq K_1\exp\ \nu(\eta^2-\xi^2) < \infty$$

for all $\xi+i\eta$ in the strip $|\xi| \leq k$, and therefore, by Corollary 4.2, $T_{\xi+i\eta}$ is of class $C^{n+k}[\alpha,\beta]$ for $|\xi| \leq k$.

In particular, $\sigma(T_{\xi+i\eta}) \subset [\alpha,\beta]$ (cf. Proposition 1.2).

Let τ_ξ denote the $C^{n+k}[\alpha,\beta]$-o.c. for T_ξ $(|\xi| \leq k)$. An argument analogous to the preceding one shows that there exists a positive constant K_2 such that

$$||\tau_\xi(f)|| \leq K_2 \max\{||\tau_{-k}(f)||, ||\tau_k(f)||\} \tag{6.8.11}$$

for all $|\xi| \leq k$ and $f \in C^{n+k}[\alpha,\beta]$. Since $\sigma(T_{\pm k}) = \sigma(S)$ by Corollary 5.23, it follows from Proposition 4.4 that if $f \in C^{n+k}[\alpha,\beta]$ vanishes in a neighborhood of $\sigma(S)$, then $\tau_{\pm k}(f) = 0$, and therefore $\tau_\xi(f) = 0$ by (6.8.11) for $|\xi| \leq k$. In other words, τ_ξ is carried by $\sigma(S)$. If $\mu \notin \sigma(S)$, there exists $f_\mu \in C^{n+k}[\alpha,\beta]$ such that $f_\mu = 0$ in a neighborhood of μ and $f_\mu = 1$ in a neighborhood of $\sigma(S)$. Set $h_\mu(t) = f_\mu(t)(\mu-t)^{-1}$ $(t \in R)$. Then $h_\mu \in C^{n+k}[\alpha,\beta]$ and $(\mu-t)\ h_\mu(t)-1 = 0$ in a neighborhood of $\sigma(S)$. Hence

$$0 = \tau_\xi[(\mu-t)h_\mu(t)-1] = (\mu 1-T_\xi)]\tau_\xi(h_\mu)-I,$$

that is, $\mu I - T_\xi$ has the inverse $\tau_\xi(h_\mu)$, and so $\mu \notin \sigma(T_\xi)$. This proves that

$$\sigma(T_\xi) \subset \sigma(S)$$

for all $\xi \in R$ (since k was arbitrary).

Fix $\xi \in R$ and set $S' = T_\xi$. Then S' is of class C^m (for suitable m) and has real spectrum (contained in $\sigma(S)$); V is S'-Volterra, and $S' - \xi V = S$. Therefore, by the preceding conclusion with S' replacing S, we have

$$\sigma(S) = \sigma(S' - \xi V) \subset \sigma(S') = \sigma(T_\xi).$$

Hence $\sigma(T_\xi) = \sigma(S)$ for all $\xi \in R$, and since $T_{\xi+i\eta}$ is similar to T_ξ, we conclude that $\sigma(T_\zeta) = \sigma(S)$ for all $\zeta \in C$.

Theorem 6.10 can be sharpened when $n = 0$, with the help of Theorem 6.9.

6.11 <u>Theorem.</u> Let k be a non-negative integer, and let S be of class C with real spectrum. Then T_ζ is of class C^k if and only if $|Re\zeta| \le k$.

<u>Proof</u>. By Theorem 6.10, T_ζ has real spectrum, and is of class C^k if $|Re\zeta| \le k$. Suppose T_ζ is of class C^k for some $\zeta = \xi + i\eta$ with $|\xi| > k$. Since $\sigma(T_\zeta) \subset R$, it follows that $||\exp(itT_{\xi+i\eta})|| = O(|t|^k)$ as $|t| \to \infty$, hence

$$(1+|t|\ ||V||)^{-|\xi|}\,||\exp(itT_{\xi+i\eta})|| = O(|t|^{k-|\xi|}) = o(1),$$

contradicting Theorem 6.9 (since $||e^{itS}|| = O(1)$ is valid for S of class C with real spectrum).

6.12 <u>Corollary</u>. Let S,V satisfy the Standing Hypothesis 6.3 in Hilbert space, and suppose $||e^{itS}|| = O(1)$. Let k be a non-negative integer. Then T_ζ is of class C^k if and only if $|Re\zeta| \le k$.

<u>Proof.</u> The condition $||e^{itS}|| = O(1)$ implies that S is similar to a self-adjoint operator (cf. proof of Theorem 2.1.1), and satisfies therefore the hypothesis of Theorem 6.11.

As a sequel to Theorem 6.10 and Corollary 5.24, the C^{n+k}-operational calculus for T_ζ is given below.

6.13 <u>Theorem</u>. Let n,k be non-negative integers. Suppose S is of class C^n with real spectrum, and let τ_0 be its $C^n[\alpha,\beta]$-o.c. Then, for $|Re\zeta| \le k$ and $f \in C^{n+k}[\alpha,\beta]$, $\mathcal{D}_{\zeta-k}\ (=\mathcal{D}V(\zeta-k) = RV(k-\zeta))$ is invariant under

$$\tau_{-k}(f) = \sum_{j=0}^{k} (-1)^j \binom{k}{j} V^j \tau_0(f^{(j)}) \qquad \text{(cf. (5.24.1))},$$

and the $C^{n+k}[\alpha,\beta]$-operational calculus $\tau_{-\zeta}$ for $T_{-\zeta}$ is given by

(1) $\tau_{-\zeta}(f) = V(\zeta-k)\tau_{-k}(f)V(k-\zeta), \quad f \in C^{n+k}[\alpha,\beta].$

For $0 \leq \mathrm{Re}\zeta \leq k$, $\mathcal{D}_{-\zeta} = RV(\zeta)$ is invariant under $\tau_0(f)$ and the expression (1) simplifies to

(2) $\tau_{\zeta}(f) = V(-\zeta)\tau_0(f)V(\zeta), \quad f \in C^{n+k}[\alpha,\beta].$

Proof. For $|\mathrm{Re}\zeta| \leq k$, $V(k-\zeta)$ is bounded, and therefore the relation

$$\tau_{-k}(f)V(k-\zeta) = V(k-\zeta)\tau_{-\zeta}(f),$$

which is true for polynomials f by Corollary 6.5, remains valid for all $f \in C^{n+k}[\alpha,\beta]$, since both T_{ζ} and T_k are of class $C^{n+k}[\alpha,\beta]$, by Theorem 6.10. Therefore $\mathcal{D}_{\zeta-k} = RV(k-\zeta)$ is invariant under $\tau_{-k}(f)$ and (1) is valid for all $f \in C^{n+k}[\alpha,\beta]$.

Suppose now that $0 \leq \mathrm{Re}\zeta \leq k$. Since $V(\zeta)$ is bounded and $T_{\pm\zeta}$ are of class $C^{n+k}[\alpha,\beta]$, the relations

$$\tau_0(f)V(\zeta) = V(\zeta)\tau_{\zeta}(f) \tag{6.13.1}$$

and $$\tau_{-\zeta}(f)V(\zeta) = V(\zeta)\tau_0(f) \tag{6.13.2},$$

which are valid for polynomials by Theorem 6.4, remain valid for all $f \in C^{n+k}[\alpha,\beta]$. In particular, $\mathcal{D}_{-\zeta} = RV(\zeta)$ is invariant under $\tau_0(f)$ by (6.13.1), and by (6.13.2) with $\zeta = k$

$$\tau_{-k}(f)V(k) = V(k)\tau_0(f).$$

Thus, by (1) with ζ replaced by $-\zeta$ $(0 \leq \mathrm{Re}\zeta \leq k)$,

$$\begin{aligned}\tau_{\zeta}(f) &= V(-\zeta-k)\tau_{-k}(f)V(k)V(\zeta)\\ &= V(-\zeta-k)V(k)\tau_0(f)V(\zeta)\\ &= V(-\zeta)\tau_0(f)V(\zeta), \qquad f \in C^{n+k}[\alpha,\beta].\end{aligned}$$

7. Convolution operators in L^p

7.0. In order to apply the abstract results of Section 6 to the classical family $M+\zeta J$ and to other concrete families of operators, we need some general tools from the theory of convolution operators. For simplicity, we shall restrict ourselves to convolutions in R, although everything is valid in R^n as well, mutatis mutandis.

As usual, S stands for the Schwartz space of rapidly decreasing functions on R. Its dual S' is the space of tempered distributions. The action of $F \in S'$ on $u \in S$ is denoted by $\langle F,u\rangle$, while $F*u$ stands for the convolution of F and u.

The Fourier transform

$$F: u \to \hat{u}(\xi) = \int_R e^{-2\pi i \xi x} u(x)\,dx$$

is a continuous automorphism of S onto itself. For $F \in S'$, the Fourier transform $\hat{F}$ is defined by the relation

$$\langle \hat{F},u\rangle = \langle F,\hat{u}\rangle \qquad (u \in S).$$

This definition coincides with the usual one when F is given by an element of $L^1 = L^1(R)$ (that is $\langle F,u\rangle = \int_R F(x)u(x)\,dx$, $u \in S$).

Let

$$S_R = \{\xi \in R;\ R/2 < |\xi| < 2R\} \qquad (R > 0)$$

and let χ_R be the characteristic function of S_R. Denote by D the differentiation operator in S'. Suppose $F \in S'$ is such that

$$\hat{F} \in L^\infty, \quad D\hat{F} \in L^2_{loc}. \tag{7.0.1}$$

when all L^p-spaces here are the usual Lebesgue spaces on R with respect to Lebesgue measure m, and L^2_{loc} stands for the space of locally square integrable functions on R.

Set

$$||F||_M = \max\{\sup_{R>0} R^{-\frac{1}{2}}||\hat{F}\cdot\chi_R||_2,\ \sup_{R>0} R^{\frac{1}{2}}||D\hat{F}\cdot\chi_R||_2\} \tag{7.0.2}$$

and let S_M' be the set of all $F \in S'$ satisfying (7.0.1) for which $||F||_M < \infty$. Clearly, $||\cdot||_M$ is a norm on S_M'. Moreover, S_M' contains the set

$$S_0' = \{F \in S';\ \hat{F}, MD\hat{F} \in L^\infty\} \qquad (7.0.3)$$

and for $F \in S_0'$,

$$||F||_M \leq \sqrt{3}\ \max\{||\hat{F}||_\infty, ||MD\hat{F}||_\infty \qquad (7.0.4)$$

where M denotes (in the expression $MD\hat{F}$) the operator of multiplication by the independent variable. Indeed,

$$R^{-\frac{1}{2}}||\hat{F}\chi_R||_2 \leq R^{-\frac{1}{2}}||\hat{F}||_\infty\ m(S_R)^{\frac{1}{2}} = \sqrt{3}||\hat{F}||_\infty$$

and

$$R^{\frac{1}{2}}||D\hat{F}\chi_R||_2 \leq R^{\frac{1}{2}}||MD\hat{F}||_\infty (2\int_{R/2}^{2R} x^{-2}dx)^{\frac{1}{2}} = \sqrt{3}||MD\hat{F}||_\infty .$$

7.1 <u>Theorem</u>. Let $F \in S_M'$. Then, for each $1 < p < \infty$,

$$||F*u||_p \leq C_p||F||_M||u||_p \qquad (u \in S)$$

where C_p depends only on p, and is bounded as $p \to \infty$.

<u>Proof</u>. The proof will involve several lemmas, that will be stated and proved when needed.

First, by homogeneity, we may assume that

$$||F||_M \leq 1 \qquad (7.1.1)$$

Fix $0 \leq \psi \in C_0^\infty$ with support in S_1 such that

$$\psi(\xi) = 1 \quad 3/4 \leq |\xi| \leq 3/2 .$$

Given $\xi \neq 0$, $\psi(\xi/2^j) \neq 0$ for $\log_2|\xi|-1 < j < \log_2|\xi|+1$ (if at all) and $\psi(\xi/2^j) = 1$ for $j = [\log_2(|\xi|/3)]+2$. Therefore

$$1 \leq \sum_{j=-\infty}^{\infty} \psi(\xi/2^j) < \infty$$

(at most two non-vanishing summands are included in the sum).

Set $\varphi(\xi) = \psi(\xi)/\sum_{j=-\infty}^{\infty} \psi(\xi/2^j)$.

Then $\varphi \in C_0^\infty$, $\mathrm{supp}\varphi \subset S_1$, and

$$\sum_{j=-\infty}^{\infty}\varphi(\xi/2^j) = 1 \qquad \text{for} \quad \xi \neq 0. \qquad (7.1.2)$$

Write $\hat{F} = f$, and

$$f_j(\xi) = f(\xi)\varphi(\xi/2^j).$$

By (7.1.2), supp $f_j \subset S_{2^j}$ and

$$f = \sum_{j=-\infty}^{\infty} f_j \qquad (\xi \neq 0). \tag{7.1.3}$$

Let $C_1 = \max\{\sup|\varphi|, \sup|\varphi'|\}$.

Since supp $f_j \subset S_{2^j}$ and $||F||_M \leq 1$, we have

$$||f_j||_2^{\,2} = ||f_j\chi_{2^j}||_2^{\,2} \leq C_1^{\,2}||f\chi_{2^j}||_2^{\,2} \leq C_1^{\,2}\, 2^j.$$

Also $|f_j'| \leq C_1(|f'| + 2^{-j}|f|)$ implies

$$||f_j'||_2^{\,2} = ||f_j'\chi_{2^j}||_2^{\,2} \leq C_1^{\,2}||(|f'| + 2^{-j}|f|)\chi_{2^j}||_2^{\,2}$$

$$\leq C_1^{\,2}(||f'\chi_{2^j}||_2 + 2^{-j}||f\chi_{2^j}||_2)^2 \leq 4C_1^{\,2}\, 2^{-j}.$$

Let g_j denote the inverse Fourier transform of f_j:

$$g_j(x) = \int e^{2\pi i x\xi} f_j(\xi)\, d\xi.$$

Then

$$-2\pi i x g_j(x) = \int e^{2\pi i x \xi} f_j'(\xi)\, d\xi,$$

and Parseval's identity implies that

$$\int (1 + 2^{2j}x^2)|g_j(x)|^2 dx = ||f_j||_2^2 + 2^{2j}(4\pi^2)^{-1}||f_j'||_2^{\,2} \leq C_2^{\,2}\, 2^j \tag{7.1.3'}$$

where $C_2 = C_1(1 + \pi^{-2})^{\frac{1}{2}}$.

Therefore, by Schwarz' inequality

$$||g_j||_1 \leq \left(\int (1+2^{2j}x^2)|g_j(x)|^2 dx\right)^{\frac{1}{2}}\left(\int (1+2^{2j}x^2)^{-1} dx\right)^{\frac{1}{2}} \leq C_2\left(\int \frac{d(2^j x)}{1+(2^j x)^2}\right)^{\frac{1}{2}} = C_2\sqrt{\pi}.$$

Hence $|f_j| = |\hat{g}_j| \leq ||g_j||_1 \leq C_2\sqrt{\pi}$ for all j. (7.1.4)

For each $\xi \neq 0$, we remarked above that $f_j(\xi) \neq 0$ for at most two indices j.

Setting

$$F_N = \sum_{j=-N}^{N} f_j, \qquad G_N = \sum_{j=-N}^{N} g_j \tag{7.1.5}$$

(so that $F_N = \hat{G}_N$), we have

$$|F_N| \leq 2C_2\sqrt{\pi} = C_3, \tag{7.1.6}$$

and by Parseval's identity, for all $u \in \mathcal{S}$,

$$||G_N * u||_2 = ||F_N\hat{u}||_2 \leq C_3||u||_2. \tag{7.1.7}$$

This inequality will be extended now to L^p-norms.

7.2 __Lemma.__ Let $1 < p < \infty$. Then for G_N as above, there exists a constant $C_p > 0$ depending only on p such that, for all $N \geq 1$,

$$||G*u||_p \leq C_p||u||_p \qquad (u \in S)$$

Assuming Lemma 7.2, we complete the proof of Theorem 7.1 as follows. By (7.1.3), (7.1.5) and (7.1.6), $F_N \to f$ pointwise (for $\xi \neq 0$) boundedly. Therefore, for $u,v \in S$, the Lebesgue dominated convergence theorem implies (since $F_N = \hat{G}_N$):

$$\begin{aligned}<F*u,v> &= <F^{-1}fFu,v> = <fFu,F^{-1}v> \\ &= \lim_{N\to\infty}<F_NFu,F^{-1}v> \\ &= \lim_{N\to\infty}<G_N*u,v>.\end{aligned}$$

By Lemma 7.2, it follows that

$$|<F*u,v>| \leq C_p||u||_p||v||_{p'}, \qquad (u,v \in S),$$

where $1/p + 1/p' = 1$.

Hence $||F*u||_p \leq C_p||u||_p \qquad (u \in S)$,

as wanted.

We turn to the proof of Lemma 7.2. Since the case $p = 2$ of the Lemma is already verified (cf. (7.1.7), it is natural to apply the following special case of the Marcinkiewicz' interpolation theorem.

7.3 __Lemma.__ Let T be a (sub)-linear operator from C_0^∞ to measurable functions on R. Suppose there exists a constant $C > 0$ such that, for all $u \in C_0^\infty$ and $\sigma > 0$,

$$m\{x \in R; |(Tu)(x)| > \sigma\} \leq (C||u||_k/\sigma)^k \tag{7.3.1}$$

for $k = 1$ and $k = 2$. Then, for all $1 < p < \infty$,

$$||Tu||_p \leq K_pC||u||_p \qquad (u \in C_0^\infty), \tag{7.3.2}$$

where $K_p > 0$ depends only on p, and is a bounded function of p for p bounded away from 1 and 2.

__Proof.__ See [49], Theorem 4.6, p.112.

Condition (7.3.1) is referred to by saying that T is of weak type (k,k) with norm $\leq$ C. Strong type (k,k) with norm $\leq$ C means that

$$||Tu||_k \leq C||u||_k \qquad (u \in C_0^\infty).$$

Clearly, strong type (k,k) with norm $\leq$ C implies weak type (k,k) with norm $\leq$ C, since

$$m\{x \in R;\ |(Tu)(x)| > \sigma\} \leq \sigma^{-k} \int_{\{x,|(Tu)(x)|>\sigma\}} |Tu|^k\, dx \leq \sigma^{-k}||Tu||_k^k \leq (C||u||_k/\sigma)^k,$$

$$(u \in C_0^\infty,\ \sigma > 0).$$

Lemma 7.3 states that weak types (1,1) and (2,2) (with norms $\leq$ C) imply strong type (p,p) for all $1 < p < \infty$ (with norm $\leq K_pC$).

Since $u \to G_N * u$ is of strong type (2,2) with norm $\leq C_3$ by (7.1.7), the proof of Lemma 7.2 (and hence of Theorem 7.1) will be completed by applying the following

7.4 Lemma. $u \to G_N * u$ is of weak type (1,1), with norm $\leq C_4$ (C_4 independent of N).

Simply take $C = \max(C_3, C_4)$ and $C_p = K_pC$. The proof of Lemma 7.4 will be based on the following general

7.5 Lemma. Let h be locally integrable on R such that, for all $t > 0$,

$$\int_{|x|\geq 2} |h^t(x-y) - h^t(x)|\,dx \leq K \quad \text{for} \quad |y| \leq 1 \qquad (7.5.1),$$

where $h^t(x) = th(tx)$.

Suppose $||h*u||_2 \leq C||u||_2 \qquad (u \in S)$. (7.5.2).

Then, for all $u \in L^1$ with compact support and $\sigma > 0$,

$$m\{x; |(h*u)(x)| > \sigma\} \leq C'||u||_1/\sigma$$

where C' depends only on K and C.

Since G_N satisfies (7.5.2) with $C = C_3$ independent of N (7.1.7), Lemma 7.4 will follow from Lemma 7.5 if we show that G_N satisfies (7.5.1) with K independent of N. Equivalently (through the change of variables $x = t\xi$, $y = t\eta$), we must verify that

$$\int_{|x|\geq 2t} |G_N(x-y) - G_N(x)|\,dx \leq K \quad \text{if} \quad |y| \leq t \qquad (7.5.3)$$

for all $N \geq 1$ and $t > 0$.

Since $\operatorname{supp} f_j \subset S_{2^j} \subset \{\xi\,; |\xi| \leq 2^{j+1}\}$, the trivial part of the Paley-Wiener theorem implies that $g_j = F^{-1}f_j$ is entire of exponential type $\leq 2^{j+1}$. By (7.1.4)

$$|g_j| \leq ||f_j||_1 = \int_{S_{2^j}} |f_j|\,d\xi \leq C_2\sqrt{\pi}\, m(S_{2^j}) = 3C_2\sqrt{\pi}\, 2^j.$$

Thus, g_j satisfies the hypothesis of the well-known Bernstein theorem (cf. [49], Theorem 7.32, p.277).

Bernstein theorem. Let h be entire of exponential type $\leq \nu$ and bounded on R. Then

$$||h'||_1 \leq \nu||h||_1.$$

In our case, by (7.1.4)

$$||g_j'||_1 \leq 2^{j+1}||g_j||_2 \leq C_2\sqrt{\pi}2^{j+1}.$$

Therefore, for $t > 0$ and $|y| \leq t$,

$$\int_{|x|\geq 2t} |g_j(x-y)-g_j(x)|dx = \int_{|x|\geq 2t} \left|\int_x^{x-y} g_j'(s)ds\right|dx$$

$$\leq \int_{-\infty}^{\infty} \left|\int_{s+y}^{s} dx\right| |g_j'(s)|ds = |y|\ ||g_j'||_1 \leq C_2\sqrt{\pi}\ 2^{j+1}t.$$

The same integral can be estimated by

$$\int_{|x|\geq 2t} |g_j(x-y)|dx + \int_{|x|\geq 2t} |g_j(x)|dx \leq 2\int_{|\xi|\geq t} |g_j(\xi)|d\xi,$$

since $\{x;|x| \geq 2t\} \subset \{\xi = x-y|\ |\xi| \geq t\}$ when $|y| \leq t$.

However, by Schwarz' inequality and (7.1.3'),

$$\int_{|x|\geq t} |g_j(x)|dx \leq (\int(1+2^{2j}x^2)|g_j(x)|^2dx)^{\frac{1}{2}}\left(\int_{|x|\geq t} \frac{dx}{2^{2j}x^2}\right)^{\frac{1}{2}}$$

$$\leq C_2 2^{j/2}\ 2^{-j}(2/t)^{\frac{1}{2}} < C_2(2^jt)^{-\frac{1}{2}}.$$

Hence

$$\int_{|x|\geq 2t} |g_j(x-y)-g_j(x)|dx \leq C_3 \min\{2^jt,\ (2^jt)^{-\frac{1}{2}}\},$$

and so, for all $t > 0$, $|y| \leq t$, and $N \geq 1$,

$$\int_{|x|\geq 2t} |G_N(x-y)-G_N(x)|dx \leq C_3 \sum_{j=-\infty}^{\infty} \min\{2^jt,\ (2^jt)^{-\frac{1}{2}}\}$$

$$= C_3\Big(t \sum_{\{j|2^jt\leq(2^jt)^{-\frac{1}{2}}\}} 2^j + (1/\sqrt{t}) \sum_{\{j|2^jt>(2^jt)^{-\frac{1}{2}}\}} (1/\sqrt{2})^j\Big)$$

$$= C_3\Big(t \sum_{k\geq\log_2 t} 1/2^k + (1/\sqrt{t}) \sum_{j>\log_2(1/t)} (1/\sqrt{2})^j\Big)$$

$$\leq C_3(2 + \sqrt{2}/(\sqrt{2} - 1)) = K,$$

and (7.5.3) is correct.

It remains to prove Lemma 7.5.

We show first that Condition (7.5.1) implies a norm estimate for $h*u$ when $\langle u,1\rangle = 0$.

7.6 Lemma. Let h be locally integrable on R and satisfy Condition (7.5.1). Then, for each $t > 0$ and $a \in R$

$$\int_{|x-a|\geq 2t} |h*u|dx \leq K||u||_1 \tag{7.6.1}$$

if $u \in L^1$ has support in $\{x; |x-a|<t\}$ and $\langle u,1\rangle = 0$.

Proof. Let u be as in the lemma. Since $\langle u,1\rangle = 0$,

$$(h^t*u)(x) = \int_R (h^t(x-y)-h^t(x))u(y)dy.$$

Let $v \to v_a$ denote the translation map,

$v_a(x) = v(a+x)$. Since $\langle (u_a)^t,1\rangle = \langle u,1\rangle = 0$,

$||(u_a)^t||_1 = ||u||_1$, $(u_a)^t(y) = 0$ for $|y| \geq 1$,

and $(h*u)_a = h*u_a$, we have

$$\int_{|x-a|\geq 2t} |(h*u)(x)|dx = \int_{|s|\geq 2t} |(h*u_a)(s)|ds =$$

$$= \int_{|\xi|\geq 2} |h^t*(u_a)^t|d\xi \leq \int_{|\xi|\geq 2}\int_{|\eta|\leq 1} |h^t(\xi-\eta)-h^t(\xi)|\ |(u_a)^t(\eta)|d\eta d\xi$$

$$= \int_{|\eta|\leq 1}\int_{|\xi|\geq 2} |h^t(\xi-\eta)-h^t(\xi)|d\xi|(u_a)^t(\eta)|d\eta$$

$$\leq K||(u_a)^t||_1 = K||u||_1.$$

Now, given any $u \in L^1$, we wish to decompose it as a sum of elements $u_k \in L^1$ with compact supports, such that $\langle u_k,1\rangle = 0$, and there is adequate control on the norms $||u_k||_1$ and on the remainder $u-\Sigma u_k$. This is the following special case of the Calderon-Zygmund "covering lemma".

7.7 Lemma. Let $u \in L^1(R)$ and let s be a fixed positive number. Then there exist disjoint open intervals I_k, and

$u_k, v \in L^1(R)$ $(k = 1,2,\ldots)$

such that

(1) $\Sigma m(I_k) \leq ||u||_1/s$;

(2) $|v| \leq 2s$ a.e. on R ;

(3) $\text{supp } u_k \subset I_k$ and $\langle u_k, 1\rangle = 0$;

(4) $||v||_1 + \Sigma||u_k||_1 \leq 3||u||_1$; and

(5) $u = v + \Sigma u_k$.

<u>Proof</u>. Subdivide R into intervals of length $> ||u||_1/s$.
For each such interval I,

$$m(I)^{-1} \int_I |u| dx < s. \tag{7.7.1}$$

Divide each I into two equal subintervals. If the mean value of $|u|$ were $\geq s$ on both halves, it would be $\geq s$ on I, contradicting (7.7.1). Let then I_{1j} be the <u>open</u> halves of the intervals I on which the mean value of $|u|$ is $\geq s$. We have by (7.7.1)

$$sm(I_{1j}) \leq \int_{I_{1j}} |u| dx \leq \int_I |u| dx < sm(I) = 2sm(I_{1j}). \tag{7.7.2}$$

Let

$$v(x) = m(I_{1j})^{-1} \int_{I_{1j}} u dy \qquad (x \in I_{1j}) \tag{7.7.3}$$

and

$$u_{1j}(x) = \begin{cases} u(x) - v(x) & (x \in I_{1j}) \\ 0 & (x \notin I_{1j}) \end{cases} \tag{7.7.4}$$

Divide in halves the subintervals of the I-s not included in the sequence $\{I_{1j}\}$, and select the <u>open</u> halves $\{I_{2j}\}$ over which the mean value of $|u|$ is $\geq s$. Extend the definition of v to $x \in I_{2j}$, using formula (7.7.3) with I_{2j} replacing I_{1j}, and define u_{2j} by (7.7.4) with the same change. Continuing this process, we obtain a sequence $\{I_k\}$ of open intervals, an associated sequence of measurable functions u_k defined on R, and a measurable function v defined on $\Omega = \cup I_k$, which we extend to R by setting $v = u$ on Ω^c.

The mean value of $|u|$ on each I_k is $\geq s$; therefore

$$s\sum m(I_k) \leq \sum \int_{I_k} |u| dx = \int_\Omega |u| dx \leq ||u||_1,$$

and (1) is verified.

If $x \in \Omega = \cup I_k$, then for precisely one k,

$$|v(x)| \leq m(I_k)^{-1} \int_{I_k} |u| dx \leq 2s$$

by property (7.7.2), valid for all the intervals I_k. If $x \notin \Omega$, there is a sequence of open intervals J_k containing x, over which the mean value of $|u|$ is $< s$, such that $m(J_k) \to 0$. Hence $|u(x)| \leq s$ for almost all x in Ω^c (cf. [43], Section 8.7). We conclude that (2) holds.

By construction, supp $u_k \subset I_k$ (cf. (7.7.4)), and $\langle u_k, 1\rangle = \int_{I_k} udx - \int_{I_k} vdx = 0$, which is (3). Since I_k are disjoint, supp $u_k \subset I_k$, and $v = u$ on Ω^c, we have (recall $u_k = u-v$ on I_k):

$$||v||_1 + \sum ||u_k||_1 = \int_{\Omega^c} |v| dx + \sum_k \int_{I_k} (|v|+|u_k|) dx$$

$$\leq \int_{\Omega^c} |u| dx + \sum 2 \int_{I_k} |v| dx + \int_{I_k} |u| dx \quad .$$

By (7.7.3), $\int_{I_k} |v| dx = |\int_{I_k} udx| \leq \int_{I_k} |u| dx$, and (4) follows.

Finally (5) is true by construction.

<u>Proof of Lemma 7.5.</u>

Let $u \in L^1$ with compact support, $||u||_1 = 1$, and decompose it as in Lemma 7.7 (with $s > 0$ to be determined):

$$u = v + \sum u_k.$$

Note that v has compact support by construction (as well as all u_k). Therefore

$$|h*u| \leq |h*v| + \sum_k |h*u_k| \tag{7.8.1}$$

By Lemma 7.6 (applied to u_k, with $I_k = \{x; |x-a_k| < \delta_k\}$; cf. Lemma 7.7(3)), we have

$$\int_{|x-a_k| \geq 2\delta_k} |h*u_k| dx \leq K ||u_k||_1 . \tag{7.8.2}$$

Let $\sum = \bigcup_k \{x; |x-a_k| < 2\delta_k\}$. By Lemma 7.7(1),

$$m(\sum) \leq 2 \sum_k m(I_k) \leq 2/s. \tag{7.8.3}$$

Next, by (7.8.2) and Lemma 7.7(4), for $\sigma > 0$,

$$m\{x \in \textstyle\sum^c;\ \sum_k |h*u_k| > \sigma/2\} \le (2/\sigma) \int_{\sum^c} \sum_k |h*u_k|dx$$

$$= (2/\sigma) \sum_k \int_{\bigcap_k\{x;|x-a_k|\ge 2\delta_k\}} |h*u_k|dx \le (2/\sigma) K\sum ||u_k||_1 \le 6K/\sigma .$$

By (7.8.3),

$$m\{x;\ \sum|h*u_k| > \sigma/2\} \le m\{x \in \textstyle\sum^c;\ \sum|h*u_k| > \sigma/2\} + m(\sum) \le 6K/\sigma + 2/s. \quad (7.8.4)$$

Since $v \in L^1 \cap L^\infty$ (cf. Lemma 7.7(2)), $v \in L^2$; hence, by (7.5.2) and the density of S in L^2,

$$||h*v||_2^{\,2} \le C^2||v||_2^{\,2} \le C^2||v||_\infty||v||_1 \le 6C^2 s,$$

where we used (2) and (4) in Lemma 7.7.

Therefore

$$m\{x;|h*v| > \sigma/2\} \le (4/\sigma^2)||h*v||_2^{\,2} \le 24C^2 s/\sigma^2. \quad (7.8.5)$$

By (7.8.1),

$$\{x;|h*u| > \sigma\} \subset \{x;|h*v| > \sigma/2\} \cup \{x;\sum_k|h*u_k| > \sigma/2\}.$$

Hence, by (7.8.4) and (7.8.5),

$$m\{x;|h*u| > \sigma\} \le 6K/\sigma + 2/s + 24C^2 s/\sigma^2.$$

The left side being independent of $s > 0$, we may minimize the right hand side with respect to s by choosing $s = \sigma/C\sqrt{12}$; whence

$$m\{x;|h*u| > \sigma\} \le C'/\sigma \quad (7.8.6)$$

with $C' = 8\sqrt{3}C + 6K$.

Finally, the assumption $||u||_1 = 1$ is removed by observing that Lemma 7.5 is trivial when $||u||_1 = 0$, and for $||u||_1 \ne 0$, we may apply (7.8.6) to $u/||u||_1$ in place of u; thus

$$m\{x;|h*u| > \sigma\} = m\{x;\ |h*(u/||u||_1)| > \sigma/||u||_1\} \le C'/(\sigma/||u||_1) = C'||u||_1/\sigma.$$

This completes the proof of Lemma 7.5 (and, therefore, of Theorem 7.1).

8. Some regular semigroups

We shall apply Theorem 7.1 to prove the regularity of certain semigroups in $L^p(0,N)$ related to the classical Volterra operator J. This will allow us among other things to apply the results of Section 6 to the family of operators $T_\zeta = M + \zeta J$ in $L^p(0,N)$ and in $C[0,N]$.

Notation. For $0 < N \leq \infty$, we let $||\cdot||_{p,N}$ denote the $L^p(0,N)$-norm, while $||\cdot||_p$ stands for the $L^p(R)$-norm. The same notation is used for the operator-norm relative to any of these spaces, whenever distinction is needed.

8.1 Definition. For $\varepsilon \geq 0$, $\zeta \in C^+$, and f locally integrable in $(0,N)$, we set

$$J_\varepsilon^\zeta f(x) = \Gamma(\zeta)^{-1} \int_0^x e^{-\varepsilon(x-t)}(x-t)^{\zeta-1} f(t)dt \qquad (x \in [0,N]).$$

The operator $J_0^\zeta = J^\zeta$ is the classical Riemann-Liouville fractional integration operator.

8.2 Lemma. (1) For each $\varepsilon \geq 0$, $\zeta \in C^+$, $1 \leq p \leq \infty$, and $0 < N < \infty$, J_ε^ζ is a bounded operator in $L^p(0,N)$ and in $C[0,N]$. The same is true in $L^p(0,\infty)$ for $\varepsilon > 0$.

(2) For $\varepsilon > 0$, $1 < p < \infty$, and $0 < N \leq \infty$,

$$||J_\varepsilon^\zeta||_{p,N} \leq C_p\, e^{\pi|\eta|/2}\, \varepsilon^{-\xi} \max\{|\zeta|,1\} \qquad (\zeta = \xi + i\eta)$$

where C_p depends only on p and is bounded as $p \to \infty$.

The same estimate (with an adequate constant C) is valid for J_ε^ζ acting in $C[0,N]$, $N < \infty$.

Proof. Let

$$K_\varepsilon^\zeta(x) = \Gamma(\zeta)^{-1} e^{-\varepsilon x} x^{\zeta-1} \qquad x > 0$$
$$= 0 \qquad x \leq 0 \tag{8.2.1}$$

$(\zeta \in C^+,\ \varepsilon \geq 0)$.

For any function f defined a.e. on $(0,N)$, set

$$f_0(x) = f(x) \qquad \text{on } (0,N)$$
$$= 0 \qquad x \notin (0,N) \tag{8.2.2}$$

Then

$$(J_\varepsilon^\zeta f)_0 = K_\varepsilon^\zeta * f_0 \tag{8.2.3}$$

for f locally integrable.

For $\varepsilon > 0$,

$$||K_\varepsilon^\zeta||_1 = |\Gamma(\zeta)|^{-1}\Gamma(\xi)\,\varepsilon^{-\xi} \qquad (\zeta = \xi + i\eta), \tag{8.2.4}$$

which implies part (1) for $\varepsilon > 0$ in view of (8.2.3).

The case $\varepsilon = 0$, $N < \infty$ is a consequence of (8.2.3) and

$$||K_0^\zeta||_{1,N} = (\xi|\Gamma(\zeta)|)^{-1}N^{\xi}. \tag{8.2.5}$$

Part (2) of the lemma is proved by applying Theorem 7.1. For $\varepsilon > 0$,

$$\hat{K}_\varepsilon^\zeta(y) = \Gamma(\zeta)^{-1}\int_0^\infty e^{-2\pi i yx-\varepsilon x}\,x^{\zeta-1}dx$$

$$= (\varepsilon^2 + 4\pi^2y^2)^{-\zeta/2}e^{-i\zeta\arctan(2\pi y/\varepsilon)}$$

(cf. [7], p.12). Differentiating, we get

$$MD\hat{K}_\varepsilon^\zeta(y) = -i\zeta(\varepsilon^2 + 4\pi^2y^2)^{-\zeta/2}\,2\pi y(\varepsilon + 2\pi i y)^{-1}e^{-i\zeta\arctan(2\pi y/\varepsilon)}$$

$$= -\zeta\,\frac{2\pi i y}{\varepsilon+2\pi i y}\,\hat{K}_\varepsilon^\zeta(y).$$

Hence

$$|MD\hat{K}_\varepsilon^\zeta(y)| \le |\zeta|\,|\hat{K}_\varepsilon^\zeta(y)| \qquad (y \in R)$$

and

$$|\hat{K}_\varepsilon^\zeta(y)| = (\varepsilon^2+4\pi^2y^2)^{-\xi/2}\,e^{\eta\arctan(2\pi y/\varepsilon)} \le e^{\pi|\eta|/2}\,\varepsilon^{-\xi} \qquad (y \in R).$$

Now, by (8.2.3), (7.0.4), and Theorem 7.1, we have for each $\varepsilon > 0$, $1 < p < \infty$, $0 < N \le \infty$, and $f \in L^p(0,N)$

$$||J_\varepsilon^\zeta f||_{p,N} = ||K_\varepsilon^\zeta * f_0||_{p,N} \le ||K_\varepsilon^\zeta * f_0||_p$$

$$\le C_p||K_\varepsilon^\zeta||_M||f_0||_p$$

$$\le C_p\sqrt{3}\,e^{\pi|\eta|/2}\varepsilon^{-\xi}\,\max\{|\zeta|,1\}\,||f||_{p,N}$$

$(\zeta = \xi+i\eta)$, where C_p denotes here the constant of Theorem 7.1.

The final statement of the lemma follows by letting $p \to \infty$, since $C_p = O(1)$ as $p \to \infty$ and $||h||_{p,N}$ tends to the $C[0,N]$-norm of h for $h \in C[0,N]$ $(0 < N < \infty)$.

8.3 Theorem. Let ε, N satisfy either $\varepsilon > 0$ and $0 < N \leq \infty$ or $\varepsilon \geq 0$ and $0 < N < \infty$. Then:

(1) For $1 \leq p < \infty$, $\{J_\varepsilon^\zeta;\ \zeta \in C^+\}$ is a holomorphic semigroup of class C_0 in $L^p(0,N)$;

(2) For $1 < p < \infty$, $\{J_\varepsilon^\zeta;\ \zeta \in C^+\}$ is a regular semigroup in $L^p(0,N)$;

(3) For $N < \infty$, $\{J_\varepsilon^\zeta;\ \zeta \in C^+\}$ is a regular semigroup in $C[0,N]$;

(4) Fix $\frac{\pi}{2} < \nu < \pi$. The boundary group $\{J_\varepsilon^{i\eta};\ \eta \in R\}$ in Cases (2) and (3) satisfies

$$||J_\varepsilon^{i\eta}|| \leq C\, e^{\nu|\eta|} \qquad (\eta \in R)$$

where C depends only on ν and p (in Case (2)), but not on N or ε.

Proof. For $1 \leq p < \infty$ and fixed $\varepsilon > 0$, the semigroup property follows from Fubini's theorem, and the C_0-property may be proved by noting that $\{\varepsilon^\xi K_\varepsilon^\xi;\ \xi > 0\}$ is an "approximate identity". The case $C[0,N]$ for $\varepsilon > 0$ and $N < \infty$ follows in a similar fashion, while the case $\varepsilon = 0$ and $N < \infty$ (for L^p or C) is well-known.

If $N < \infty$, one has for $\zeta, \omega \in C^+$

$$||J_\varepsilon^\zeta f - J_\varepsilon^\omega f|| \leq ||J^\zeta(e^{\varepsilon t} f) - J^\omega(e^{\varepsilon t} f)|| \tag{8.3.1}$$

for the $L^p(0,N)$ and the $C[0,N]$ norms, where $J^\zeta = J_0^\zeta$ is the usual Riemann-Liouville semigroup. Since $\{J^\zeta;\ \zeta \in C^+\}$ is known to be strongly continuous, the same follows from (8.3.1) for $\{J_\varepsilon^\zeta;\ \zeta \in C^+\}$, $\varepsilon > 0$.

For the case $L^p(0,\infty)$, $\varepsilon > 0$, we use (8.3.1) for the $L^p(0,N)$-norm, $0 < N < \infty$ arbitrary; thus

$$||J_\varepsilon^\zeta f - J_\varepsilon^\omega f||_{p,\infty}^p \leq ||J^\zeta(e^{\varepsilon t} f) - J^\omega(e^{\varepsilon t} f)||_{p,N}^p$$

$$+ \int_N^\infty e^{-xp\varepsilon/2} |\Gamma(\zeta)^{-1} \int_0^x (x-t)^{\zeta-1}\, e^{\varepsilon(t-x)/2}\, e^{\varepsilon t/2} f(t)dt|^p\, dx$$

$$+ \int_N^\infty e^{-xp\varepsilon/2} |\Gamma(\omega)^{-1} \int_0^x (x-t)^{\omega-1}\, e^{\varepsilon(t-x)/2} e^{\varepsilon t/2} f(t)dt|^p\, dx.$$

Let E denote the set of functions f measurable on $(0,\infty)$, such that $e^{\varepsilon t/2} f(t) \in L^p(0,\infty)$. If $f \in E$, then by (8.2.4), we can estimate the sum of the two integrals over (N,∞) by

$$F(\zeta,\omega)^p e^{-N\varepsilon p/2}||e^{\varepsilon t/2}f(t)||_{p,\infty}^p$$

where

$$F(\zeta,\omega) = |\Gamma(\zeta)|^{-1}\Gamma(\xi)(\varepsilon/2)^{-\xi} + |\Gamma(\omega)|^{-1}\Gamma(\sigma)(\varepsilon/2)^{-\sigma} \qquad (\sigma = \text{Re } \omega).$$

For $\zeta \in C^+$ fixed and ω belonging to some closed disc centered at ζ and lying in C^+, we have $F(\zeta,\omega) \leq M$ for some constant $M > 0$. Hence, given $\delta > 0$, $F(\zeta,\omega)e^{-N\varepsilon/2}||e^{\varepsilon t/2}f||_{p,\infty} < \delta$ for a suitable $N > 0$. For this N, apply the strong continuity of J^ζ in $L^p(0,N)$ to obtain

$$\limsup_{\omega\to\zeta}||J_\varepsilon^{\zeta}f - J_\varepsilon^{\omega}f||_{p,\infty} \leq \delta \qquad (f \in E).$$

Since δ is arbitrary, $J_\varepsilon^{\omega}f \to J_\varepsilon^{\zeta}f$ as $\omega \to \zeta$ (in $L^p(0,\infty)$), for each $f \in E$. However, since E is dense in $L^p(0,\infty)$ and $||J_\varepsilon^{\zeta}||_{p,\infty}$ is bounded on compact subsets of C^+ (cf. (8.2.4)), the strong continuity of $\{J_\varepsilon^{\zeta};\ \zeta \in C^+\}$ in $L^p(0,\infty)$ follows. The analyticity is then a consequence of Morera's theorem.

For $\varepsilon > 0$, $1 < p < \infty$, and $0 < N \leq \infty$, we apply Lemma 8.2(2) to obtain

$$\sup_Q ||J_\varepsilon^{\zeta}||_{p,N} \leq C_p\sqrt{2}e^{\pi/2}\max\{\varepsilon^{-1},1\}, \tag{8.3.2}$$

with a similar estimate for operator norms of J_ε^{ζ} over $C[0,N]$, $N < \infty$. Here Q denotes the rectangle $\{\zeta \in C;\ 0 < \xi \leq 1,\ |\eta| \leq 1\}$. Thus $\{J_\varepsilon^{\zeta};\ \zeta \in C^+\}$ is a regular semigroup in $L^p(0,N)$ $(1 < p < \infty,\ 0 < N \leq \infty)$ and in $C[0, N]$ $(N < \infty)$ when $\varepsilon > 0$. Let $\{J_\varepsilon^{i\eta};\ \eta \in R\}$ denote its boundary group (cf. Theorem 6.1). Letting $\xi \to 0+$ in Lemma 8.2(2), we get

$$||J_\varepsilon^{i\eta}||_{p,N} \leq \limsup_{\xi\to 0+}||J_\varepsilon^{\xi+i\eta}||_{p,N} \leq C_p e^{\pi|\eta|/2}\max\{|\eta|,1\}.$$

Fixing $\frac{\pi}{2} < \nu < \pi$, we then have

$$||J_\varepsilon^{i\eta}||_{p,N} \leq C\, e^{\nu|\eta|} \qquad (\eta \in R) \tag{8.3.3}$$

where C depends only on p and ν, but not on N or ε $(\varepsilon > 0,\ N \leq \infty)$. A similar estimate is valid over $C[0,N]$, $N < \infty$.

Next, fix $N < \infty$. Since $0 \leq 1-e^{\varepsilon(t-x)} \leq \varepsilon(x-t)$ for $0 \leq t \leq x$, we have

$$|(J^{\xi+i\eta}f - J_\varepsilon^{\xi+i\eta}f)(x)| \leq \varepsilon|\Gamma(\xi+i\eta)|^{-1}\ \Gamma(\xi+1)(J^{\xi+1}|f|)(x).$$

Since $||J^{\xi+1}||_{p,N} \leq \Gamma(\xi+2)^{-1}N^{\xi+1}$, it follows that

$$||J^{\xi+i\eta}f - J_\varepsilon^{\xi+i\eta}f||_{p,N} \leq \varepsilon(|\xi+i\eta|N^{\xi+1}/(\xi+1)|\Gamma(\xi+1+i\eta)|)||f||_{p,N}$$

$$\leq \varepsilon M||f||_{p,N} \qquad (8.3.4)$$

for $\xi+i\eta \in Q = \{\xi+i\eta;\ 0 < \xi \leq 1,\ |\eta| \leq 1\}$, where M denotes the supremum in Q of the function $|\xi+i\eta|N^{\xi+1}/(\xi+1)|\Gamma(\xi+1+i\eta)|$ (which is obviously continuous in $\overline{Q}$, hence bounded in Q).

We write

$$||J^{\xi+i\eta}f||_{p,N} \leq ||J^{\xi+i\eta}f - J_\varepsilon^{\xi+i\eta}f||_{p,N} + ||J_\varepsilon^{\xi+i\eta}f||_{p,N} \qquad (8.3.5)$$

and apply (8.3.2) and (8.3.4) with $\varepsilon = 1$; thus

$$\sup_Q ||J^{\xi+i\eta}||_{p,N} \leq M'$$

where $M' = C_p\sqrt{2}e^{\pi/2} + M$, and similarly for operator norms over $C[0,N]$.

This shows that $\{J^\zeta;\ \zeta \in C^+\}$ is regular over $L^p(0,N)$ $(1 < p < \infty)$ and over $C[0,N]$.

Let then $\{J^{i\eta};\ \eta \in R\}$ be the boundary group. Fix $\nu \in (\frac{\pi}{2}, \pi)$. For any $\varepsilon > 0$ and $f \in L^p(0,N)$, we have by (8.3.3), (8.3.4), and (8.3.5):

$$||J^{i\eta}f||_{p,N} = \lim_{\xi\to 0+} ||J^{\xi+i\eta}f||_{p,N} \leq [\varepsilon|\eta|N/|\Gamma(1+i\eta)| + Ce^{\nu|\eta|}]\ ||f||_{p,N},$$

where C depends only on p and ν (but <u>not</u> on ε). Letting $\varepsilon \to 0$, we obtain

$$||J^{i\eta}f||_{p,N} \leq Ce^{\nu|\eta|}\ ||f||_{p,N} \qquad (\eta \in R),$$

and similarly for $C[0,N]$. This completes the proof of Theorem 8.3.

We consider now the families of operators

$$T_{\zeta,\varepsilon} = M + \zeta J_\varepsilon \qquad (\zeta \in C,\ \varepsilon \geq 0)$$

in $L^p(0,N)$ or $C[0,N]$, $0 < N < \infty$, $1 < p < \infty$. Since $[M,J_\varepsilon] = J_\varepsilon^{\,2}$, the Standing Hypothesis of Section 6.3 is satisfied by the semigroup $V(\zeta) = J_\varepsilon^{\,\zeta}$ $(\zeta \in C^+)$ (cf. Theorem 8.3), with $B = M$. In our case, B is of class C with spectrum equal to $[0,N]$ (cf. Section 2.2). Thus, all the results of Section 6 are valid in the present situation, including Theorem 6.11, which we restate as follows.

8.4 <u>Theorem.</u> Let $0 < N < \infty$, $1 < p < \infty$, and $\varepsilon \geq 0$. Consider the operators $T_{\zeta,\varepsilon} = M + \zeta J_\varepsilon$ in $L^p(0,N)$ or $C[0,N]$, for $\zeta \in \mathbb{C}$.

Then $\sigma(T_{\zeta,\varepsilon}) = [0,N]$, and $T_{\zeta,\varepsilon}$ is of class C^k if and only if $|\mathrm{Re}\,\zeta| \leq k$.

Actually, Theorem 8.4 is equivalent to its special case with $\varepsilon = 0$. Indeed, if $M_\varphi : f(x) \to \varphi(x)f(x)$, then for $\varphi(x) = e^{\varepsilon x}$, $J_\varepsilon = M_\varphi^{-1} J M_\varphi$, and consequently $T_{\zeta,\varepsilon} = M_\varphi^{-1} T_\zeta M_\varphi$ where $T_\zeta = M + \zeta J$, since M_φ commutes with M. Since both spectrum and C^k-classification are similarity invariants, our claim is evident. In fact, for any function φ such that $\varphi, 1/\varphi \in L^\infty(0,N)$ (resp., $\varphi \in C[0,N]$, $\varphi \neq 0$ in $[0,N]$), the operator $f(x) \to xf(x) + \zeta\int_0^x [\varphi(t)/\varphi(x)]f(t)dt$ in $L^p(0,N)$ (resp., $C[0,N]$) has spectrum $[0,N]$, and is of class C^k if and only if $|\mathrm{Re}\,\zeta| \leq k$.

9. Similarity

Let $T = S + \zeta V$ $(\zeta \in C)$, where the pair (S,V) satisfies the Standing Hypothesis 6.3. We saw that T_ζ is similar to T_ω if $\operatorname{Re}\zeta = \operatorname{Re}\omega$ (cf. Corollary 6.5). In case $\|e^{itS}\| = O(1)$, this can be sharpened as follows.

9.1 Theorem. Let S,V satisfy the Standing Hypothesis, and suppose that $\|e^{itS}\| = O(1)$. Then, for $\zeta,\omega \in C$, the operators T_ζ and T_ω are similar if $\operatorname{Re}\zeta = \operatorname{Re}\omega$ and only if $|\operatorname{Re}\zeta| = |\operatorname{Re}\omega|$.

Proof. The sufficiency of the condition $\operatorname{Re}\zeta = \operatorname{Re}\omega$ is included in Corollary 6.5. The necessity part is a special case of the following

9.2 Theorem. Let (S_k,V_k) $(k = 1,2)$ be two pairs of operators satisfying the Standing Hypothesis, and suppose $\|\exp(itS_k)\| = O(1)$ $(k = 1,2)$. Then, for $\zeta,\omega \in C$ with $|\operatorname{Re}\zeta| \neq |\operatorname{Re}\omega|$, $S_1 + \zeta V_1$ is not similar to $S_2 + \omega V_2$.

Proof. Set $T_{\zeta,k} = S_k + \zeta V_k$ $(k = 1,2)$.
By Corollary 6.5, it suffices to show that the operators $T_{\xi,1}$ and $T_{\lambda,2}$ are not similar for real ξ,λ with $|\xi| > |\lambda|$. Suppose they are, that is $T_{\xi,1} = Q^{-1}T_{\lambda,2}Q$ for a non-singular operator Q. Then $\exp(itT_{\xi,1}) = Q^{-1}\exp(itT_{\lambda,2})Q$ for all $t \in R$. Since $\|\exp(itS_k)\| = O(1)$, we may apply Theorem 6.9 to the pair (S_1,V_1) (let $C_1(\cdot)$ be the function associated to this pair as in Theorem 6.9), and then use Theorem 6.8 with the pair (S_2,V_2). Hence

$$0 < C_1(\xi) \le (1+|t|\,\|V_1\|)^{-|\xi|}\|\exp(itT_{\xi,1})\|$$

$$\le K\|Q\|\,\|Q^{-1}\|(1+|t|\,\|V_1\|)^{-|\xi|}(1+|t|\,\|V_2\|)^{|\lambda|} \to 0 \text{ as } |t| \to \infty;$$

contradiction.

9.3 Corollary. Let $T_\zeta = M + \zeta J$ $(\zeta \in C)$, acting in $L^p(0,N)$ $(1 < p < \infty,\ 0 < N < \infty)$ or $C[0,N]$ $(0 < N < \infty)$. Then T_ζ is similar to T_ω if and only if $\operatorname{Re}\zeta = \operatorname{Re}\omega$.

Proof. The pair (M,J) satisfies the Standing Hypothesis (cf. remarks preceding Theorem 8.4), and $\|e^{itM}\| = 1$. Therefore Theorem 9.1 will imply our corollary if we prove that T_ξ and $T_{-\xi}$ are not similar for ξ real $\neq 0$. Suppose they are similar, that is $T_\xi Q = QT_{-\xi}$ with Q non-singular. By Theorem 6.4

$$M(J^\xi QJ^\xi) = J^\xi(T_\xi Q)J^\xi = J^\xi(QT_{-\xi})J^\xi = (J^\xi QJ^\xi)M,$$

that is $A = J^{\xi}QJ^{\xi}$ commutes with M, and is therefore a multiplication operator $M_{\varphi}: f(x) \to \varphi(x)f(x)$. (Indeed, for any polynomial q,

$Aq = Aq(M))1 = q(M)A1 = M_{\varphi}q$ with $\varphi = A1$, and therefore $A = M_{\varphi}$ by density of polynomials in $C[0,N]$ or $L^p(0,N)$; $\varphi = A1 \in C[0,N]$ in the first case, and the relation $M_{\varphi} = A \in B(L^p(0,N))$ implies that $\varphi \in L^{\infty}(0,N)$ in the second case.) Hence

$$\sigma(A) = \overline{\text{range } \varphi} \text{ (or } \overline{\text{essential range } \varphi}). \qquad (9.3.1)$$

Since $\xi \neq 0$, J^{ξ} is compact; hence $A = J^{\xi}QJ^{\xi}$ is compact. In the case of $C[0,N]$, φ is continuous, and therefore its range contains an interval (unless φ is constant). However (9.3.1) and the discreteness of the spectrum of compact operators exclude this possibility, that is $\varphi = c$ (a constant) and $A = cI$. By compactness of A, we must have $A = J^{\xi}QJ^{\xi} = 0$, which is impossible because J^{ξ} is one-to-one (cf. Theorem 6.2) and Q is non-singular. A general argument, valid for $L^p(0,N)$ as well, goes as follows. Let λ be a non-zero eigenvalue of A, and let $f \neq 0$ be a corresponding eigenvector. For $n = 1,2,\ldots$, $AM^n f = M^n Af = \lambda M^n f$, that is the λ-eigenspace of A contains the span of $\{M^n f\}$, which is infinite dimensional, since $f \neq 0$. But this is impossible, by compactness of A. Hence $\sigma(A) = \{0\}$, and therefore $\varphi = 0$ by (9.3.1); that is $A = 0$, which is impossible , as we saw before.

9.4 <u>Corollary</u>. Let $0 < N < \infty$ and $\varphi, 1/\varphi \in L^{\infty}(0,N)$ (or $\varphi \in C[0,N]$ and $\varphi \neq 0$ in $[0,N]$). Consider the operator

$$(V_{\varphi}f)(x) = \int_0^x [\varphi(t)/\varphi(x)]f(t)dt$$

acting in $L^p(0,N)$, $1 < p < \infty$ (resp., in $C[0,N]$). Let ψ satisfy the same hypothesis as φ. Then $M + \zeta V_{\varphi}$ and $M + \omega V_{\psi}$ $(\zeta, \omega \in C)$ are similar if and only if $\mathrm{Re}\,\zeta = \mathrm{Re}\,\omega$.

<u>Proof</u>. Since $M + \zeta V_{\varphi} = M_{\varphi}^{-1}(M + \zeta J)M_{\varphi}$ with $M_{\varphi}: f(x) \to \varphi(x)f(x)$, the result follows trivially from Corollary 9.3.

In connection with Corollary 9.3, we may investigate similarity within the wider family $\{M + \zeta J^{\alpha};\ \zeta, \alpha \in C;\ \mathrm{Re}\,\alpha \geq 0\}$. For $\mathrm{Re}\,\alpha \geq 1$, $\alpha \neq 1$, the solution of this problem turns out to be a simple consequence of Lemma 5.2 and Theorem 8.3. We first prove a general result.

9.5 Proposition. Let A be a Banach algebra with identity over $\mathbb{C}$. Let $s,v \in A$ be such that $c = [s,v]$ commutes with v. Then for all $\zeta \in \mathbb{C}$, $s + \zeta c$ is similar to s with $e^{\zeta v}$ intertwining

$$s + \zeta c = e^{-\zeta v} s e^{\zeta v} \qquad (\zeta \in \mathbb{C}). \tag{9.5.1}$$

Conversely, if for some $c,v \in A$ and all t in some real interval, $s + tc$ is similar to s with e^{tv} intertwining, then $c = [s,v]$ and c commutes with v.

Proof. Take $D = D_s: x \to [s,x]$ and $f:\lambda \to e^{\zeta\lambda}$ ($\zeta \in \mathbb{C}$ fixed) in Lemma 5.2. This give

$$[s,e^{\zeta v}] = \zeta e^{\zeta v} c$$

which is equivalent to (9.5.1).

Conversely, suppose $e^{tv}(s + tc) = se^{tv}$ for some $s,v,c \in A$ and all t in some real interval. Expanding in power series and comparing coefficients of t^k, we get $[s,v^k] = kv^{k-1}c$ $(k = 1,2,\ldots)$. For $k = 1$, this establishes that $c = [s,v]$. Using then the relation for $k = 2$, we obtain $cv = sv^2 - vsv = v^2s + 2vc - vsv = vc$.

9.6 Consider for example the elements $s = M$ and $v = J^{\alpha-1}/(\alpha-1)$ (for $\alpha \in \mathbb{C}$, $\mathrm{Re}\alpha \geq 1$, $\alpha \neq 1$, in $A = B(L^p(0,N))$ $(1 < p < \infty$ if $\mathrm{Re}\alpha \geq 1$ or $1 \leq p < \infty$ if $\mathrm{Re}\alpha > 1)$ or in $A = B(C[0,N])$. Since $c = [s,v] = [M,J^{\alpha-1}]/(\alpha-1) = J^\alpha$ (cf. Theorem 8.3) commutes with v, it follows that <u>$M + \zeta J^\alpha$ is similar to M for all $\zeta \in \mathbb{C}$</u>, with $e^{\zeta J^{\alpha-1}/(\alpha-1)}$ intertwining. This situation contrasts drastically with that of Corollary 9.3 (the case $\alpha = 1$), according to which <u>$M + \zeta J$ is similar to M if and only if $\mathrm{Re}\zeta = 0$</u>.

The similarity problem for the family $M + \zeta J^\alpha$ for α in the strip $0 \leq \mathrm{Re}\alpha < 1$ remains open. Does the situation for $\mathrm{Re}\alpha \geq 1$ $(\alpha \neq 1)$ "extend analytically" to the strip $0 \leq \mathrm{Re}\alpha < 1$, or is there a "propagation" of the singular behavior at $\alpha = 1$?

9.7 Corollary. Let s,v,c be as in Proposition 9.5, and let g be a complex function analytic in a neighborhood of $\sigma(v)$. Then $s + g'(v)c$ is similar to s with $e^{g(v)}$ intertwining.

Proof. By Lemma 5.2, $[s,g(v)] = g'(v)c$ commutes with $g(v)$, so that we may apply Proposition 9.5 with $g(v)$ and $g'(v)c$ replacing v and c respectively.

9.8 Corollary. Let A be a complex Banach algebra with identity. Let $s,v_j \in A$ be such that v_j and $c_j = [s,v_j]$ commute with v_i $(1 \leq i,j \leq k)$. Let g_i be complex functions analytic in a neighborhood of $\sigma(v_i)$, $1 \le i \le k$. Then $s + \sum_{i=1}^{k} g_i'(v_i)c_i$ is similar to s, with $\exp(\sum_{i=1}^{k} g_i(v_i))$ intertwining.

Proof. For $1 \le j \le k$, set

$$s_j = s + \sum_{i=1}^{j} g_i'(v_i)c_i \tag{9.8.1}$$

and write $s_0 = s$. Clearly $[s_j, v_{j+1}] = c_{j+1}$ $(j < k)$ commutes wtih v_{j+1}. Applying Corollary 9.7 with $g = g_{j+1}$ we obtain for $0 \le j < k$:

$$s_{j+1} = s_j + g_{j+1}'(v_{j+1})c_{j+1} = \exp(-g_{j+1}(v_{j+1}))s_j\exp(g_{j+1}(v_{j+1})).$$

This recursion formula implies the Corollary.

9.9 Corollary. Let A be a complex Banach algebra with identity. Let $s, v_j \in A$ be such that v_j and $c_j = [s, v_j]$ commute with v_i $(1 \le i,j < \infty)$. Let g_i be complex functions analytic in a neighborhood of $\sigma(v_i)$ $(i = 1,2,\ldots)$, such that $\sum_{i=1}^{\infty} g_i(v_i)$ converge in A to an element w. Then $s + \sum_{i=1}^{\infty} g_i'(v_i)c_i = s_\infty$ converge in A, and is similar to s with e^w intertwining:

$$s_\infty = e^{-w}se^{w}.$$

Proof. Let $w_j = \sum_{i=1}^{j} g_i(v_i)$ $(j = 1,2,\ldots)$. Then $\exp(\pm w_j) \to \exp(\pm w)$, and therefore, by Corollary 9.8 (cf. notation (9.8.1)):

$$s_j = \exp(-w_j)s\,\exp(w_j) \to \exp(-w)s\,\exp(w).$$

The following result generalizes the example in Section 9.6.

9.10 Corollary. Let $S, V(\cdot)$ satisfy the Standing Hypothesis 6.3. Let $\alpha_i, \zeta_i \in C$ $(\mathrm{Re}\,\alpha_i \ge 1,\ \alpha_i \ne 1)$ be such that $W = \sum_{i=1}^{\infty} \zeta_i V(\alpha_i - 1)/(\alpha_i - 1)$ converge in $B(X)$. Then $S + \sum_{i=1} \zeta_i V(\alpha_i)$ converges in $B(X)$ and is similar to S with e^W intertwining.

Proof. In the Banach algebra $A = B(X)$, take $s = S$ and $v_j = V(\alpha_j - 1)/(\alpha_j - 1)$. By Theorem 6.4, $c_j = [s, v_j] = V(\alpha_j)$, so that v_j and c_j commute with v_i $(i,j = 1,2,\ldots)$, and the result follows from Corollary 9.9 with $g_i(\lambda) = \zeta_i\lambda$.

9.11 Corollary. Let $S, V(\cdot)$ satisfy the Standing Hypothesis 6.3. Let $\alpha_i, \zeta_i \in C$ be as in Corollary 9.10, and set

$$T_\zeta = S + \zeta V + \sum_{i=1}^{\infty} \zeta_i V(\alpha_i) \qquad (\zeta \in \mathbb{C}).$$

Then:

(1) if S is of class C^n with real spectrum, then $\sigma(T_\zeta) = \sigma(S)$, and T_ζ is of class C^{n+k} for ζ in the strip $|\mathrm{Re}\zeta| \leq k$.

(2) if S is of class C with real spectrum, then T_ζ is of class C^k if and only if $|\mathrm{Re}\zeta| \leq k$.

(3) T_ζ is similar to T_ω if $\mathrm{Re}\zeta = \mathrm{Re}\omega$.

(4) if $||e^{itS}|| = O(1)$, T_ζ is not similar to T_ω when $|\mathrm{Re}\zeta| \neq |\mathrm{Re}\omega|$.

Proof. Set $S' = S + \sum_{i=1}^{\infty} \zeta_i V(\alpha_i)$. Since $[S',V] = [S,V] = V^2$ and S' is similar to S by Corollary 9.10, the result follows from Theorem 6.10, Theorem 6.11, Corollary 6.5, and Theorem 9.1 (in this order).

In particular, for α_i, ζ_i as above and

$$T_\zeta = M + \zeta J + \sum_{i=1}^{\infty} \zeta_i J^{\alpha_i} \qquad (\zeta \in \mathbb{C}),$$

operating in $L^p(0,N)$ $(1 < p < \infty)$ or $C[0,N]$, T_ζ is of class C^k iff $|\mathrm{Re}\zeta| \leq k$, and T_ζ is similar to T_ω iff $\mathrm{Re}\zeta = \mathrm{Re}\omega$.

Another interesting special case of Corollary 9.11 is obtained by taking $\alpha_i = i+1$. Recalling that V is quasi-nilpotent (cf. Lemma 5.13), the convergence condition on W is certainly satisfied when ζ_i are the power series coefficients of a function f analytic at 0.

9.12 Corollary. Let $S, V(\cdot)$ satisfy the Standing Hypothesis 6.3, and let f be analytic at 0 with $f(0)$ real. Then

(1) if S is of class C^n with real spectrum, then $\sigma(S + f(V)) = \sigma(S) + f(0)$ and $S + f(V)$ is of class C^{n+k} for $|\mathrm{Re}f'(0)| \leq k$.

(2) if S is of class C with real spectrum, then $S + f(V)$ is of class C^k if and only if $|\mathrm{Re}f'(0)| \leq k$.

(3) Let f,g be analytic at 0 with $f(0) = g(0)$ real and $\mathrm{Re}f'(0) = \mathrm{Re}g'(0)$. Then $S + f(V)$ is similar to $S + g(V)$. Conversely,

(4) If $||e^{itS}|| = O(1)$ and $S + f(V)$ is similar to $S + g(V)$, then $f(0) = g(0)$ and $|\mathrm{Re}f'(0)| = |\mathrm{Re}g'(0)|$.

In particular, in $L^p(0,N)$ $(1 < p < \infty)$ or $C[0,N]$, the following generalization of Theorem 8.4 and Corollary 9.3 is valid.

9.13 <u>Corollary</u>. (1) Let f be analytic at 0 with $f(0)$ real. Then $M + f(J)$ is of class C^k if and only if $|\mathrm{Re}f'(0)| \leq k$.

(2) Let f,g be analytic at 0 with $f(0)$ and $g(0)$ real. Then $M + f(J)$ is similar to $M + g(J)$ if and only if $f(0) = g(0)$ and $\mathrm{Re}f'(0) = \mathrm{Re}g'(0)$.

9.14 The behavior of the family of operators $\{M + \zeta J^\alpha;\ \alpha,\zeta \in C,\ \mathrm{Re}\alpha > 1\}$ as regard to similarity (cf §9.6) can be put in perspective by considering perturbations of M by convolution operators.

Let Δ be a finite interval. For $1 \leq p < \infty$, let $L^p(\Delta)$ denote the usual L^p-space, considered as a subspace of $L^p(R)$:

$$L^p(\Delta) = \{f \in L^p(R);\ f = 0 \text{ a.e outside } \Delta\}.$$

A <u>weight</u> on Δ is a measurable function $w\colon \Delta \to (0,\infty)$ satisfying $w(x+y) \leq w(x)w(y)$ for $x,y,x+y \in \Delta$.

Let $L_w(\Delta)$ denote the space of all complex measurable functions h on Δ such that $wh \in L^1(\Delta)$. Set

$$C_h f(x) = \int_\Delta \frac{w(x)}{w(t)} h(x-t) f(t)\,dt,\ x \in \Delta$$

for $f \in L^p(\Delta)$, with $h \in L_w(\Delta)$ fixed.

Since all functions involved are understood to vanish outside Δ, integration extends over the interval $\max(\alpha, x-\beta) \leq t \leq \min(x-\alpha,\beta)$ where $\Delta = (\alpha,\beta)$. In particular, both t and $x-t$ are in Δ (for $x \in \Delta$), so that $\frac{w(x)}{w(t)} \leq w(x-t)$, and

$$|C_h f| \leq (w|h|) * |f| \quad \text{on } \Delta,$$

hence
$$||C_h f||_p \leq ||wh||_1 ||f||_p, \quad \text{that is} \quad ||C_h|| \leq ||wh||_1.$$

Suppose h is a complex measurable function on Δ such that $g(x) = h(x)/x \in L_w(\Delta)$. If $K = \sup_{x\in\Delta}|x|$, then $|h(x)| \leq K|g(x)|$, so that $h \in L_w(\Delta)$ as well. We are now ready to state the following corollary of Proposition 9.5, where the operators are understood to act in $L^p(\Delta)$, $1 \leq p < \infty$.

9.15 <u>Corollary</u>. Suppose $g(x) = h(x)/x \in L_w(\Delta)$. Then $M + C_h$ is similar to M, and the most general intertwining operator Q is of the form $Q = M_\varphi \exp(C_g)$, with φ and $1/\varphi$ in $L_\infty(\Delta)$.

<u>Proof.</u> In the Banach algebra $A = B(L^p(\Delta))$, take $s = M$ and $v = C_g$. Then $c = [s,v] = C_h$ commutes with $v = C_g$, since any pair of convolution operators (with the same weight) commute. By Proposition 9.5, $M + C_h$ similar to M with $P = \exp(C_g)$ intertwining. If Q is an arbitrary intertwining operator for $M + C_h$ and M, then $M + C_h = P^{-1}MP = Q^{-1}MQ$, that is $(QP^{-1})M = M(QP^{-1})$. Therefore QP^{-1} is a multiplication operator M_φ with $\varphi \in L^\infty(\Delta)$ (cf. proof of Corollary 9.3). Since $M_{1/\varphi} = (QP^{-1})^{-1} \in B(L^p(\Delta))$, one has also $1/\varphi \in L^\infty(\Delta)$.

For example, if $\Delta = (0,N)$ with $0 < N < \infty$, $w(x) \equiv 1$, and $h(x) = \zeta x^{\alpha-1}/\Gamma(\alpha)$ ($\alpha,\zeta \in \mathbb{C}$, $\mathrm{Re}\,\alpha > 0$), then $g(x) = h(x)/x \in L_w(\Delta) = L^1(\Delta)$ if and only if $\mathrm{Re}\,\alpha > 1$ (when $\zeta \neq 0$). Thus $M + \zeta J^\alpha$ is similar to M for all $\zeta,\alpha \in \mathbb{C}$, $\mathrm{Re}\,\alpha > 1$.

<u>Remark.</u> The same proof gives a "vector-valued version" of Corollary 9.15. Let X be a Banach space, and denote by $L^p(\Delta,X)$ the Banach space of all (equivalence classes of) measurable functions $f\colon \Delta \to X$ such that $||f|| \in L^p(\Delta)$. Given a weight w on Δ and $h\colon \Delta \to B(X)$ such that $wh \in L^1(\Delta,B(X))$, set

$$(C_h f)(x) = \int_\Delta \frac{w(x)}{w(t)} h(x-t) f(t)\,dt, \qquad x \in \Delta$$

for $f \in L^p(\Delta,X)$.

If $h(x)$ commutes with $h(y)$ for all $x,y \in \Delta$ and if $w(x)h(x)/x \in L^1(\Delta,B(X))$, we obtain as before that $M + C_h$ is similar to M, where $M\colon f(x) \to xf(x)$ in $L^p(\Delta,X)$.

Similarly, if D is a bounded domain in $\mathbb{C}$ and $M\colon f(z) \to zf(z)$ in $L^p(D,X)$ (with respect to plane Lebesgue measure), we may consider perturbations $M + C_h$ with "weighted convolution operators" C_h defined as follows: Let $w\colon D \to (0,\infty)$ be a weight function ($w(z+\zeta) \le w(z)w(\zeta)$ whenever $z,\zeta,z+\zeta \in D$), and let

$$(C_h f)(z) = \int_D \frac{w(z)}{w(\zeta)} h(z-\zeta) f(\zeta)\,d\xi\,d\eta,$$

where $\zeta = \xi + i\eta$, $f \in L^p(D,X)$, $h \in L^q(D,B(X))$ and p,q are suitably related.

9.16 <u>Corollary.</u> Let $1 \le p < \infty$, and let D be a bounded domain in $\mathbb{C}$. Let $h\colon D \to B(X)$ be measurable with commuting range ($[h(x),h(y)] = 0$ for all $x,y \in D$). Suppose $wh \in L^q(D,B(X))$ for some $q > 4$ such that $q \ge \max(p,\frac{p}{p-1})$. Then $M + C_h$

is similar to M.

Proof. Set $A(z,\zeta) = \frac{w(z)}{W(\zeta)} h(z-\zeta)$ $(z = z + iy)$.

Then

$$\int_D \int_D ||A(z,\zeta)||^q dxdyd\xi d\eta \le \int_D \int_D [w(z-\zeta)||h(z-\zeta)||]^q dxdyd\xi d\eta \le (\text{meas } D)^q ||wh||_q^q < \infty.$$

Setting $g(x) = h(x)/x$, C_g is then an operator of the type defined in [5, p.2411, formula (42)]. Therefore C_h and C_g are bounded operators on $L^p(D,X)$, and since $[M,C_g] = C_h$ commutes with C_g (because h has a commuting range in $B(X)$), the conclusion follows from Proposition 9.5.

10. Spectral Analysis

In the preceding section, we encountered certain classes of perturbations $S + H$ that were similar to S. In particular, if S is a scalar type spectral (s.t.s) operator in Dunford's sense [5;III], the same is true of $S + H$. For example, the operator $T_\zeta = M + \zeta J$ acting in $L^p(0,N)$ $(1 < p < \infty)$ or $C[0,N]$ is similar to M for $\mathrm{Re}\zeta = 0$, and is therefore s.t.s (for $\mathrm{Re}\zeta = 0$). For $\mathrm{Re}\zeta \neq 0$, T_ζ is not of class C (by Theorem 8.4), hence not s.t.s; that is T_ζ is a scalar type spectral operator if and only if $\mathrm{Re}\zeta = 0$. We may consider the more general question: for what values of $\zeta \in \mathbb{C}$ is T_ζ spectral (not necessarily of **scalar** type). This problem (and some ramifications) will be solved in the abstract context of the Standing Hypothesis 6.3.

10.1 Spectral operators. We collect in this section some standard definitions, notation, and results, referring to [5;III] for details.

Let X be a Banach space. We denote by $\mathcal{B} = \mathcal{B}(\mathbb{C})$ the σ-algebra of all Borel subsets of $\mathbb{C}$, considered as a Boolean algebra. A spectral measure on $\mathcal{B}$ is an algebra homomorphism $E: \mathcal{B} \to B(X)$ such that $x^*E(\cdot)x$ is a regular countably additive complex measure on $\mathcal{B}$, for each $x \in X$ and $x^* \in X^*$. Necessarily, E is bounded $(K = \sup_{\delta \in \mathcal{B}} ||E(\delta)|| < \infty)$ and countably additive in the strong operator topology.

We say that $T \in B(X)$ is a scalar type spectral (s.t.s) operator if there exists a spectral measure E such that

$$T = \int_{\sigma(T)} \lambda E(d\lambda)$$

where the integral is defined as usual in the strong operator topology.

For each $\delta \in \mathcal{B}$, the projection $E(\delta)$ commutes necessarily with T, so that $E(\delta)X$ is a reducing subspace for T, and one verifies easily that $\sigma(T|E(\delta)X) \subset \bar{\delta}$. This motivates the more general concept of a spectral operator: $T \in B(X)$ is spectral if there exists a spectral measure E on $\mathcal{B}$ commuting with T such that $\sigma(T|E(\delta)X) \subset \bar{\delta}$ for each $\delta \in \mathcal{B}$. The spectral measure E is uniquely determined and is called the resolution of the identity for T. The spectral operator T has a unique Jordan decompostion $T = S + N$ with S (the scalar part of T) and N (the radical part of T) commuting, S s.t.s and N quasi-nilpotent. In fact, if E is the resolution of the identity for T, then $S = \int \lambda E(d\lambda)$, with integration extended over $\sigma(S) = \sigma(T)$. Conversely, any operator of the form $T = S + N$, with S s.t.s and N quasinilpotent commuting, is spectral with resolution of the identity coinciding with the spectral measure for S. When the radical part N is nilpotent, we say that T is spectral of finite type.

An operator $T \in B(X)$ is said to have the single-valued extension property (s.v.e property) if whenever $f: D_f \to X$ is analytic in an open set $D_f \subset C$ and satisfies

$$(\lambda I - T)f(\lambda) = 0 \quad (\lambda \in D_f), \tag{10.1.1}$$

it follows that $f = 0$ in D_f.

If T has the s.v.e property, and $x \in X$, we denote by $\rho_T(x)$ the set of all $\alpha \in C$ such that there exists an X-valued function $x(\cdot)$ analytic in a neighborhood V_α of α, such that

$$(\lambda I - T)x(\lambda) = x \qquad (\lambda \in V_\alpha) \tag{10.1.2}$$

The s.v.e property insures the uniqueness of the function $x(\cdot)$, which is clearly an analytic extension of $R(\cdot;T)x$. In particular, $\rho(T) \subset \rho_T(x)$. The complement $\sigma_T(x) = C \setminus \rho_T(x)$ is the local spectrum of T at x; it is a compact subset of $\sigma(T)$, which is non-empty for $x \neq 0$ (the usual Liouville theorem argument!)

For $F \subset C$, we set

$$X_T(F) = \{x \in X;\ \sigma_T(x) \subset F\}. \tag{10.1.3}$$

This is a T-invariant linear manifold. If T is spectral, then T has the s.v.e. property, and

$$X_T(F) = E(F)X \tag{10.1.4}$$

for each closed $F \subset C$.

We shall use the notation of Sections 5.3 and 5.6 in the Banach algebra $B(X)$. Recall that $U, T \in B(X)$ are quasi-nilpotent equivalent ($U \sim T$) if $d(T,U) = 0$ (cf. Section 5.10).

10.2 Lemma. Let $T,U \in B(X)$, $T \sim U$. Then if U has the s.v.e. **property**, so does T, and $\sigma_T(x) = \sigma_U(x)$ for each $x \in X$.

Proof. Suppose $f: D_f \to X$ is analytic and

$$(\lambda I - T)f(\lambda) = 0 \qquad (\lambda \in D_f). \tag{10.2.1}$$

Fix $\lambda \in D_f$ and set

$$G_\lambda(\mu) = \sum_{j=0}^{\infty} C(U,T)^j I\cdot(\mu-\lambda)^{-j-1} \qquad (\mu \neq \lambda).$$

Since $T \sim U$, the root test shows that the series converges in $B(X)$, so that $G_\lambda(\cdot)$ is analytic in $\mathbb{C}\setminus\{\lambda\}$.

We have

$$(\mu I - U)G_\lambda(\mu) = [R_{\mu I-T} - C(U,T)]G_\lambda(\mu)$$

$$= \sum_{j=0}^{\infty} C(U,T)^j I(\mu I-T)/(\mu-\lambda)^{j+1} - \sum_{j=0}^{\infty} C(U,T)^{j+1} I/(\mu-\lambda)^{j+1}.$$

Since $(\mu I-T)f(\lambda) = (\mu-\lambda)f(\lambda)$ by (10.2.1), we obtain

$$(\mu I-U)G_\lambda(\mu)f(\lambda) = f(\lambda) \qquad (\mu \neq \lambda). \tag{10.2.2}$$

Hence $\mathbb{C}\setminus\{\lambda\} \subset \rho_U(f(\lambda))$.

Rewrite (10.2.2) with λ replaced by $\zeta \in D_f$ and use Cauchy's formula for the circle $|\zeta-\lambda| = \delta$ with δ sufficiently small.

Thus

$$(\mu I-U)\frac{1}{2\pi i}\int_{|\zeta-\lambda|=\delta} \frac{G_\zeta(\mu)f(\zeta)}{\zeta-\lambda}\,d\zeta = f(\lambda),$$

and since the integral defines a function of μ analytic in a neighborhood of λ, this shows that $\lambda \in \rho_U(f(\lambda))$. Hence $\rho_U(f(\lambda)) = \mathbb{C}$, and therefore $f(\lambda) = 0$. Since $\lambda \in D_f$ was arbitrary, this shows that T has the s.v.e. property.

Next, fix $\alpha \in \rho_T(x)$ ($x \in X$ given). Then there exists an analytic function $x(\cdot): V_\alpha \to X$ such that

$$(\lambda I-T)x(\lambda) = x \qquad (\lambda \in V_\alpha). \tag{10.2.3}$$

Deriving, we see that

$$(\lambda I-T)x^{(j)}(\lambda) = -jx^{(j-1)}(\lambda) \qquad (\lambda \in V_\alpha, j \geq 1). \tag{10.2.4}$$

Consider the formal series ■

$$y(\lambda) = \sum_{j=0}^{\infty} (-1)^j C(U,T)^j I\cdot x^{(j)}(\lambda)/j! \tag{10.2.5}$$

If the series converges in X, then

$$(\lambda I-U)y(\lambda) = \sum_{j=0}^{\infty} (-1)^j [R_{\lambda I-T} - C(U,T)]C(U,T)^j I\cdot x^{(j)}(\lambda)/j!$$

$$= \sum_{j=0}^{\infty} (-1)^j C(U,T)^j I\cdot(\lambda I-T)x^{(j)}(\lambda)/j! - \sum_{j=0}^{\infty} (-1)^j C(U,T)^{j+1} I\cdot x^{(j)}(\lambda)/j!$$

$$\equiv x$$

by (10.2.3) and (10.2.4).

If $r(U,T) = 0$ (cf. (5.6.1)), it follows from Cauchy's inequalities that the series for $y(\lambda)$ converges in a neighborhood of α. Hence $y(\cdot)$ is analytic in a neighborhood of α and $(\lambda I-U)y(\lambda) \equiv x$ there. This shows that $\alpha \in \rho_U(x)$, that is $\rho_T(x) \subset \rho_U(x)$. If also $r(T,U) = 0$ (that is $U \sim T$), we obtain $\rho_T(x) = \rho_U(x)$ by symmetry.

Note that the above argument proves the following:

10.3 <u>Lemma</u>. If $U,T \in B(X)$ have the s.v.e. property and $r(U,T) = 0$, then $\sigma_U(x) \subset \sigma_T(x)$ for each $x \in X$.

10.4 <u>Lemma</u>. Let $U,T \in B(X)$ be spectral operators such that $r(U,T) = 0$. Then $U \sim T$.

<u>Proof</u>. By Lemma 10.3, $\sigma_U(x) \subset \sigma_T(x)$ for each $x \in X$, and therefore

$$X_T(F) \subset X_U(F)$$

hence
$$E_T(F)X \subset E_U(F)X$$

for each closed set $F \subset C$ (E_T and E_U denote the resolutions of the identity for T and U respectively; cf. (10.1.4)). Hence $E_U(F)E_T(F) = E_T(F)$ (F closed). For $G \subset C$ open, write G as the union of an increasing sequence of closed **sets**, and use the strong σ-additivity of spectral measures to obtain $E_U(G)E_T(G) = E_T(G)$ (G open). Applying this to $G = F^c$ (the complement of the closed set F), we obtain

$$I - E_T(F) = E_T(F^c) = E_U(F^c)E_T(F^c)$$

$$= I - E_T(F) - E_U(F) + E_U(F)E_T(F)$$

$$= I - E_U(F)$$

that is, $E_T(F) = E_U(F)$ for all closed sets F, hence also for all open sets G, and therefore $E_T = E_U$ by regularity of the measures x^*E_Tx and x^*E_Ux ($x \in X$, $x^* \in X$). It follows that $S_T = S_U$ (where S_T and S_U denote the scalar parts of T and U respectively). By Remark 5.9(2), if a,b are commuting elements of a Banach algebra A, then $d(a,b) = r(a-b)$, where $r(x)$ denotes the spectral radius of $x \in A$. Therefore $a \sim b$ if and only if $a-b$ is quasi-nilpotent. In particular, any spectral operator $T = S + N$ is quasi-nilpotent equivalent to its scalar part. Hence $T \sim S_T = S_U \sim U$, so that $T \sim U$ by transitivity.

10.5 Lemma. Let $U,T \in B(X)$, $U \sim T$. If T is spectral, then so is U, and $E_T = E_U$.

Proof. By Lemma 10.2, U has the s.v.e. property and $E_T(F)X = X_T(F) = X_U(F)$ for each closed set F. In particular, $E_T(F)X$ is U-invariant, that is

$$E_T(F)UE_T(F) = UE_T(F) \qquad (F \text{ closed}).$$

As in the preceding proof, we obtain the same relation for each open set G (using the strong σ-additivity of E_T). Taking $G = F^c$, we get

$$\begin{aligned} U - UE_T(F) &= UE_T(F^c) = E_T(F^c)UE_T(F^c) \\ &= U - UE_T(F) - E_T(F)U + E_T(F)UE_T(F) \\ &= U - E_T(F)U, \end{aligned}$$

that is $E_T(F)U = UE_T(F)$, F closed. The same is true for open sets (by complementation), and finally $E_T(\delta)U = UE_T(\delta)$ for any $\delta \in \mathcal{B}$, by regularity of the measures x^*E_Tx ($x \in X, x^* \in X^*$).

Now, for each $\delta \in \mathcal{B}$, $E_T(\delta)X$ is a reducing subspace for U (and for T as well!). Denote $T_\delta = T|E_T(\delta)X$ and $U_\delta = U|E_T(\delta)X$, and similarly for any operator commuting with $E_T(\delta)$. We have

$$\begin{aligned} ||C(U_\delta,T_\delta)^n I_\delta||^{1/n} &= ||\sum_{j=0}^{n} (-1)^j \binom{n}{j} U_\delta^{n-j} T_\delta^{\,j}||^{1/n} \\ &= ||(\sum_{j=0}^{n} (-1)^j \binom{n}{j} U^{n-j} T^j)_\delta||^{1/n} \\ &\le ||C(U,T)^n I||^{1/n} \to 0 \end{aligned}$$

as $n \to \infty$, that is $r(U_\delta,T_\delta) = 0$, and similarly $r(T_\delta,U_\delta) = 0$. Hence $U_\delta \sim T_\delta$, and therefore, by Corollary 5.10,

$$\sigma(U_\delta) = \sigma(T_\delta) \subset \overline{\delta}, \qquad (\delta \in \mathcal{B}).$$

This proves that U is spectral, with resolution of the identity E_T. By uniqueness of the resolution of the identity, $E_U = E_T$.

10.6 Lemma. Let V be S-Volterra , and suppose that both S and S + kV are spectral for some integer $k \neq 0$. Then $S + \zeta V$ is spectral and quasinilpotent equivalent to S for all $\zeta \in C$.

Proof. By Lemma 5.13(2), either $r(S + kV,S)$ or $r(S,S + kV)$ vanishes, and therefore $S + kV \sim S$ by Lemma 10.4. By Corollary 5.15, it follows that $S + \zeta V \sim S$ for all $\zeta \in C$. Hence $S + \zeta V$ is spectral for all $\zeta \in C$, by Lemma 10.5.

The following "non-spectrality" result can be proved without the machinery of Section 6.

10.7 Proposition. Suppose $||e^{itS}|| = O(1)$ for $t \in R$, and V is a one-to-one S-Volterra operator. Then $T_n = S + nV$ is not spectral for positive integers n.

Proof. Suppose T_n is spectral for some positive integer n, and let $T_n = A + N$ be its Jordan decomposition. Since $||e^{itS}|| = O(1)$, $\sigma(S)$ is real (cf. (3.1.4)). By Corollary 5.23, $\sigma(T_n) = \sigma(S)$, hence $\sigma(A) = \sigma(S)$ is real, and therefore $||e^{itA}|| = ||\int_R e^{it\lambda}E(d\lambda)|| = O(1)$.

By Corollary 5.20, for $t \in R$,

$$e^{itN} = e^{-itA}\exp(itT_n) = e^{-itA}e^{itS}(I+itV)^n \tag{10.7.1}$$

and consequently, as $|t| \to \infty$,

$$||e^{itN}|| = O(|t|^n) \qquad (t \in R). \tag{10.7.2}$$

The entire function e^{izN} $(z \in C)$ with values in $B(X)$ is of order one and minimal type (cf. [11;p.104]), because N is quasi-nilpotent. It follows from (10.7.2) and [11; Theorem 3.13.8] that e^{izN} is a polynomial of degree $\leq n$. By uniqueness of power series expansion, this means that $N^{n+1} = 0$. Let $x \in X$; then as $|t| \to \infty$, by (10.7.1),

$$\begin{aligned}||V^nNx|| &= \lim||(\tfrac{I}{it} + V)^nNx|| \\ &= \lim|t|^{-n}||(I + itV)^nNx|| \\ &= \lim|t|^{-n}||e^{-itS}e^{itA}\sum_{k=0}^{n-1}(it)^kN^{k+1}x/k!|| \\ &\le K\lim\sum_{k=0}^{n-1}|t|^{k-n}||N^{k+1}x||/k! = 0,\end{aligned}$$

that is $V^nNx = 0$, and since V is one-to-one, $Nx = 0$. Hence $N = 0$, and therefore, by (10.7.1), $||(I + itV)^n|| = ||e^{-itS}e^{itA}|| = O(1)$. It follows that $V = 0$, contradicting the injectivity of V.

We shall use now the machinery of Section 6 to prove a similar result for $T_\zeta = S + \zeta V$ with ζ complex, when $\mathrm{Re}\zeta$ is <u>not</u> an integer.

10.8 <u>Proposition</u>. Suppose $||e^{itS}|| = O(1)$ for $t \in R$, and assume the operators S,V satisfy the Standing Hypothesis 6.3. Let $\zeta = \xi + i\eta \in C$, ξ not an integer. Then $T_\zeta = S + \zeta V$ is not spectral.

<u>Proof</u>. By Corollary 6.5, T_ζ is similar to T_ξ, and since spectrality is similarity invariant, it suffices to prove that T_ξ is not spectral. Suppose T_ξ is spectral, and let $T_\xi = A + N$ be its Jordan decomposition. Let n be the unique positive integer such that $n-1 < |\xi| < n$. As in the proof of Proposition 10.7, we see that $||e^{itA}|| = O(1)$, and therefore, by Theorem 6.8,

$$||e^{itN}|| = ||e^{-itA}\exp(itT_\xi)|| \le K(1+|t|\ ||V||)^{|\xi|} = o(|t|^n).$$

By [11; Theorem 3.13.8], it follows as in the preceding proof that e^{izN} $(z \in C)$ is a polynomial of degree $< n$, and therefore $N^n = 0$. Now, as $|t| \to \infty$,

$$\begin{aligned}&(1+|t|\ ||V||)^{-|\xi|}\ ||\exp(itT_\xi)|| = \\ &=(1+|t|\ ||V||)^{-|\xi|}\ ||e^{itA}\sum_{k=0}^{n-1}(it)^kN^k/k!|| \\ &= O(|t|^{n-1-|\xi|}) \to 0,\end{aligned}$$

contradicting Theorem 6.9.

10.9 <u>Remark</u>. By Lemma 3.2, the hypothesis $||e^{itS}|| = O(1)$ implies that S is of class C^2 (with real spectrum). Therefore, by Corollary 5.24, if V is S-Volterra, then $T_n = S + nV$ is of class $C^{|n|+2}$ for any integer n. Similarly, by Theorem 6.10, the hypothesis of Proposition 10.8 imply that T_ζ is of class

C^{n+2} for all ζ in the strip $|\mathrm{Re}\,\zeta| \leq n$ $(n = 0,1,2,\ldots)$. In Hilbert space, the condition $\|e^{itS}\| = O(1)$ is necessary and sufficient for S to be of class C with real spectrum (see Section 2.1.1). Therefore, by Theorem 6.11, the hypothesis of Proposition 10.8 imply that T_ζ is of class C^n if and only if $|\mathrm{Re}\,\zeta| \leq n$ $(n = 0,1,2,\ldots)$. These positive C^n-classification results should be compared to the negative spectrality results of Propositions 10.7 and 10.8, and also to the following

10.10 Theorem. Let S be a scalar type spectral operator with real spectrum and suppose S,V satisfy the Standing Hypothesis. Then $T_\zeta = S + \zeta V$ $(\zeta \in C)$ is spectral if and only if $\mathrm{Re}\,\zeta = 0$.

When this is the case, T_ζ is similar to S, hence of scalar type.

Proof. If $\mathrm{Re}\,\zeta = 0$, then T_ζ is similar to S by Corollary 6.5, and is therefore a spectral operator (of scalar type).

Since S is spectral of scalar type with real spectrum, $\|e^{itS}\| = O(1)$, and therefore T_ζ is not spectral whenever $\mathrm{Re}\,\zeta$ is not an integer (by Proposition 10.8).

If $\mathrm{Re}\,\zeta = k \neq 0$ is an integer, then T_ζ is similar to $T_k = S + kV$ (by Corollary 6.5), and it remains to show that T_k is not spectral. If it were, then by Lemma 10.6, T_ζ would be spectral for all $\zeta \in C$, contradicting Proposition 10.8.

10.11 Corollary. Let $S,V \in B(H)$, H a Hilbert space; suppose the Standing Hypothesis is satisfied, and $\|e^{itS}\| = O(1)$. Then $T_\zeta = S + \zeta V$ is spectral if and only if $\mathrm{Re}\,\zeta = 0$ (in which case T_ζ is similar to S, hence to a self-adjoint operator).

Proof. See Remark 10.9 and Section 2.1.1.

10.12 Corollary. Let $T_\zeta = M + \zeta J$ acting in $L^p(0,N)$, $0 < N < \infty$, $1 < p < \infty$. Then T_ζ is spectral if and only if $\mathrm{Re}\,\zeta = 0$.

Proof. See end of Section 3.3 and apply Theorem 10.10.

10.13 Corollary. Let $S,V(\cdot)$ satisfy the Standing Hypothesis 6.3. Let $\alpha_i,\zeta_i \in C$ be as in Corollary 9.10, and suppose S is a scalar type spectral operator with real spectrum. Set

$$T_\zeta = S + \zeta V + \sum_{i=1}^{\infty} \zeta_i V(\alpha_i) \qquad (\zeta \in C).$$

Then T_ζ is spectral if and only if $\mathrm{Re}\zeta = 0$.

Proof. See proof of Corollary 9.11.

In particular, for α_i, ζ_i as above, and $T_\zeta = M + \zeta J + \sum_{i=1}^{\infty} \zeta_i J^{\alpha_i}$ $(\zeta \in C)$, T_ζ is spectral iff $\mathrm{Re}\zeta = 0$ (when it acts in $L^p(0,N)$, $1 \le p < \infty$).

As in Section 9.12, we obtain the following result which is true in particular for the pair (M,J).

10.14 Corollary. Let S,V satisfy the Standing Hypothesis 6.3, with S scalar type spectral with real spectrum. Let f be a complex function analytic at 0, with $f(0)$ real. Then $S + f(V)$ is spectral if and only if $\mathrm{Re}f'(0) = 0$.

Since S is necessarily of class C (with real spectrum), Corollary 9.12(2) states that $S + f(V)$ is of class C^k if and only if $|\mathrm{Re}f'(0)| \le k$ $(k = 0,1,2,\ldots)$. This should be compared to Corollary 10.14.

11. The family $S + \zeta V$, S unbounded

Some of the results of Sections 6 and 9 will be extended to the case when S is an unbounded operator. In view of Theorem 8.3, this will enable us to analyze the family of unbounded operators $M + \zeta J_\varepsilon$ $(\varepsilon > 0,\ \zeta \in C)$ in $L^p(0,\infty)$ in a manner analogous to the one used in $L^p(0,N)$, $N < \infty$. As expected, however, the results will not be as definitive as in the case of bounded S.

The Banach algebra methods of Section 5 are not available in the present situation. We shall proceed now through the theory of semigroups, assuming as a start that iS generates a strongly continuous group of operators $S(\cdot)$ (it will replace e^{itS} in the preceding analysis). In our standard example $S = M$: $f(x) \to xf(x)$ in $L^p(0,\infty)$, we have $S(t)$: $f(x) \to e^{itx}f(x)$. Note that $D(M) = \{f \in L^p(0,\infty);\ xf(x) \in L^p(0,\infty)\}$.

11.1 Notation. The following notation will be fixed throughout this section.

Let $\{S(t); t \in R\}$ be a strongly continous group of operators in the Banach space X. Denote its infinitesimal generator by iS; S is a closed operator with domain $D(S)$ dense in X. Let $R(\lambda;S)$ be the resolvent of S, for $\lambda \in \rho(S)$ (the resolvent set of S).

11.2 Lemma. The following statements are equivalent for an operator $V \in B(X)$:

(a) $VD(S) \subset D(S)$ and $[S,V] \subset V^2$;

(b) $[R(\lambda;S),V] = R(\lambda,S)V^2R(\lambda;S)$ $(\lambda \in \rho(S))$;

(c) $[S(t),V] = itVS(t)V$ $(t \in R)$;

(d) same as (c), for t in some interval $(0,s)$.

Proof. (a) $\Rightarrow$ (b). By (a),

$$(\lambda I - S)V = V(\lambda I - S) - V^2 \quad \text{on } D(S).$$

For each $\lambda \in \rho(S)$ and $x \in X$, $R(\lambda;S)x \in D(S)$, hence $VR(\lambda;S)x \in D(S)$, and therefore

$$\begin{aligned}[R(\lambda;S),V]x &= R(\lambda;S)Vx - R(\lambda;S)[(\lambda I - S)V]R(\lambda;S)x \\ &= R(\lambda;S)Vx - R(\lambda;S)[V(\lambda I - S) - V^2]R(\lambda;S)x \\ &= R(\lambda;S)V^2R(\lambda;S)x.\end{aligned}$$

(b) ⇒ (c). For $x \in D(S)$, $t > 0$ and c large enough, we use the strongly convergent integral representation (cf. [11; p.622]):

$$S(t)x = \int_R e^{(c+is)t} R(c+is;iS)x\,ds/2\pi$$

$$= \int_R e^{i(s-ic)t} R(s-ic;S)x\,ds/2\pi i.$$

For $\lambda \in \rho(S)$ and $x \in D(S)$, it follows from (b) that

$$Vx = VR(\lambda;S)(\lambda I-S)x$$

$$= R(\lambda;S)V(I-VR(\lambda;S))(\lambda I-S)x \in D(S),$$

i.e., $VD(S) \subset D(S)$.

Therefore, for $x \in D(S)$ and $t > 0$, we have by (b):

$$[S(t),V]x = \int_R e^{i(s-ic)t}[R(s-ic;S),V]x\,ds/2\pi i$$

$$= \int_R e^{i(s-ic)t}\,R(s-ic;S)V^2R(s-ic;S)x\,ds/2\pi i.$$

Since $||R(\lambda;S)|| = O(1/|\lambda|)$ as $|\lambda| \to \infty$, we have $||VR(\lambda;S)|| < 1$ for $|\lambda|$ large enough, and therefore $[I-VR(\lambda;S)]^{-1}$ exists in $B(X)$ (for such λ). By (b),

$$VR(\lambda;S) = R(\lambda;S)V[I-VR(\lambda;S)],$$

hence

$$R(\lambda;S)V = VR(\lambda;S)[I-VR(\lambda;S)]^{-1}$$

commutes with $VR(\lambda;S)$ for $|\lambda|$ large enough (hence for all λ outside some horizontal strip). Therefore, for c large enough,

$$[S(t),V]x = V\int_R e^{i(s-ic)t}R(s-ic;S)^2Vx\,ds/2\pi i$$

$$= -V\int_R e^{i(s-ic)t}\,\frac{d}{ds}\,[R(s-ic;S)Vx]ds/2\pi i.$$

Integrating by parts, we obtain (since $R(s-ic;S) = O(1/|s|)$ as $|s| \to \infty$):

$$[S(t),V]x = itV\int_R e^{i(s-ic)t}R(s-ic;S)Vx\,ds/2\pi i$$

$$= itVS(t)Vx$$

for x in $D(S)$, and hence for all x, by continuity (since $D(S)$ is dense in

X). This proves (c) for $t > 0$. For $t = 0$, (c) is trivial. For $t < 0$, consider the semigroup $S_-(t) = S(-t)$ $(t > 0)$, with infinitesimal generator $-iS$. Since $R(\lambda;-S) = -R(-\lambda;S)$, we have by (b)

$$[R(\lambda;-S),-V] = R(\lambda;-S)(-V)^2R(\lambda;-S),$$

that is, the pair $(-S,-V)$ satisfies (b). It follows that $[S_-(t),-V] = it(-V)S_-(t)(-V)$ for $t > 0$, that is $[S(-t),V] = -itVS(-t)V$, which is property (c) for $t < 0$.

(c) $\Rightarrow$ (d); trivial.

(d) $\Rightarrow$ (a). Let $x \in D(S)$ and $t \in (0,s)$. By (d),

$$t^{-1}(S(t)-I)Vx = Vt^{-1}(S(t)-I)x + iVS(t)Vx \to V(iSx) + iV^2x \quad \text{as} \quad t \to 0.$$

Hence $Vx \in D(S)$ and $(iS)Vx = i(VSx + V^2x)$, which proves (a).

We need next an elementary algebraic lemma. Let G be an abelian (additive) group, and let H be a semigroup with identity 1. A map $f: G \to H$ is a homomorphism if $f(0) = 1$ and $f(x+y) = f(x)f(y)$ for all $x,y \in G$. A map $g: G \to H$ is "abelian" if its range is an abelian subset of H.

11.3 Lemma. Let G be an abelian group, and H a semigroup with identity. Let $f,g: G \to H$ be such that f and fg are homomorphisms and g is abelian. Then fg^r is a homomorphism for all $r \in Z$ (multiplications are defined pointwise).

Proof. Since f and $h = fg$ are homomorphisms, $f(x)$ and $h(x)$ are invertible in H, and therefore $g(x) = f(x)^{-1}h(x)$ is invertible for each $x \in G$. Since g is abelian, $g(x)^{-1}$ commutes with $g(y)^{-1}$ and with $g(y)$ for all $x,y \in G$. Let $k = fg^{-1}$. Then $k(0) = 1$, and for all $x,y \in G$, using the hypothesis about f, g, and h,

$$\begin{aligned} k(x+y)k(y)^{-1}k(x)^{-1} &= f(x)f(y)g(x+y)^{-1}g(y)f(y)^{-1}g(x)f(x)^{-1} \\ &= f(x)h(y)g(x+y)^{-1}f(y)^{-1}f(x)^{-1}h(x)f(x)^{-1} \\ &= f(x)h(y)h(x+y)^{-1}f(x+y)f(y)^{-1}f(x)^{-1}h(x)f(x)^{-1} \\ &= 1, \end{aligned}$$

that is k is a homomorphism.

Thus the pair (f,g^{-1}) satisfies the same hypothesis as the pair (f,g), and consequently, it suffices to prove the lemma for <u>positive</u> integers r. This is done by induction on r. For $r = 1$, there is nothing to prove. Assume fg^r is a homomorphism for some $r \geq 1$ (and of course for $r = 1$). Hence

$$f(y)g(x+y)^r = g(x)^r f(y)g(y)^r \qquad (x,y \in G)$$

holds for the given r and for $r = 1$.

Since g is abelian, we obtain for all $x,y \in G$:

$$f(x+y)g(x+y)^{r+1}g(y)^{-r-1}f(y)^{-1}g(x)^{-r-1}f(x)^{-1} =$$

$$f(x)g(x)^r f(y)g(y)^r g(x+y)g(y)^{-r-1}f(y)^{-1}g(x)^{-r-1}f(x)^{-1} =$$

$$f(x)g(x)^r g(x)f(y)g(y)g(y)^r g(y)^{-r-1}f(y)^{-1}g(x)^{-r-1}f(x)^{-1} = 1.$$

Since $(fg^{r+1})(0) = 1$ holds trivially, we conclude that fg^{r+1} is a homomorphism.

Set $T_\zeta = S + \zeta V$ $(\zeta \in C)$ for $V \in B(X)$ given. By [11, Theorem 13.2.2], iT_ζ is the infinitesimal generator of a strongly continuous group of operators, which we denote by $T_\zeta(\cdot)$.

11.4 <u>Lemma</u>. Let $V \in B(X)$ be such that $VD(S) \subset D(S)$ and $[S,V] \subset V^2$. Then $\rho(V)$ contains the imaginary axis without 0, and for all $k \in Z$ and $t \in R$,

$$T_k(t) = (I-itV)^{-k}S(t) = S(t)(I+itV)^k.$$

<u>Proof</u>. Suppose we proved that $T_k(t) = S(t)(I+itV)^k$, $k \in Z, t \in R$. Then

$$T_k(t) = T_k(-t)^{-1} = [S(-t)(I-itV)^k]^{-1}$$

$$= (I-itV)^{-k}S(t).$$

It suffices then to show that

$$T_k(t) = S(t)(I+itV)^k \qquad (k \in Z, t \in R).$$

Set $H(t) = S(t)(I+itV)$. Then $H(0) = I$, and by Lemma 11.2(c),

$$\begin{aligned} H(t)H(u) &= S(t)(I+itV)S(u)(I+iuV) \\ &= S(t)\{S(u)+itVS(u)+iuS(u)V+it[S(u),V]\} \\ &= S(t)S(u)\{I+i(t+u)V\} = H(t+u) \end{aligned}$$

for all $t,u \in R$.

Using the language of Lemma 11.3, the maps $t \to S(t)$ and $t \to H(t) = S(t)(I+itV)$ are homomorphisms of the additive group R into $B(X)$, and $t \to G(t) = I+itV$ is abelian. By Lemma 11.3, it follows that $H_k(t) = S(t)(I+itV)^k$ is a group, for each $k \in Z$. Clearly, $H_k(\cdot)$ is strongly continuous. Let iA_k be its infinitesimal generator. It remains to show that $A_k = T_k$.

Let $x \in D(S)$ and $k \in Z^+$. Then, for $t > 0$,

$$t^{-1}(H_k(t)x-x) = t^{-1}[S(t)-I](I+itV)^k x+t^{-1}[(I+itV)^k x-x]$$

$$= t^{-1}[S(t)-I]x+[S(t)-I](ikVx+O(t))+ikVx+O(t) \to iSx+ikVx$$

$$= iT_k x \quad \text{as} \quad t \to 0,$$

that is $A_k = T_k$ on $D(S) = D(T_k)$.

Since $I+itV = S(-t)H(t)$, $(I+itV)^{-1} \in B(X)$ for all $t \in R$, that is $\rho(V)$ contains the imaginary axis without 0.

For $t > 0$ small enough, the preceding argument is clearly valid for k negative as well, so that $T_k \subset A_k$ for all $k \in Z$.

However $S(t) = H_k(t)(I+itV)^{-k}$, and by Lemma 11.2

$$[H_k(t),V] = [S(t),V](I+itV)^k$$

$$= itVS(t)V(I+itV)^k = itVH_k(t)V$$

for all $t \in R$. By Lemma 11.2, this is equivalent to the hypothesis $VD(A_k) \subset D(A_k)$ and $[A_k,V] \subset V^2$. We may then replace the group $S(\cdot)$ by the group $H_k(\cdot)$ provided k is replaced by $-k$: the generator iS of the group $S(t) = H_k(t)(I+itV)^{-k}$ is therefore an extension of the generator $i(A_k-kV)$ (corresponding to T_k above), that is $S+kV \supset A_k$. This completes the proof of the lemma.

When S is unbounded, the hypothesis $[S,V] \subset V^2$ does not necessarily imply that V is quasi-nilpotent, so that Lemma 11.4 does not make sense in general when $k \in Z$ is replaced by $\zeta \in C$. The following slight extension of Lemma 11.4 will be sufficient for our purposes.

11.5 <u>Corollary.</u> Let $V \in B(X)$ be such that $VD(S) \subset D(S)$ and $[S,V] \subset V^2$.

Then for all $\zeta \in C$, $k \in Z$, and $t \in R$,

$$T_{\zeta+k}(t) = T_\zeta(t)(I+itV)^k = (I-itV)^{-k}T_\zeta(t).$$

Proof. Fix $\zeta \in C$. The pair (T_ζ, V) satisfies the hypothesis of Lemma 11.4 on (S,V); when S is replaced by $T_\zeta = S+\zeta V$, $T_k = S+kV$ is clearly replaced by $T_\zeta + kV = T_{\zeta+k}$.

When t is restricted so that $(I+itV)^\zeta$ makes sense as a convergent binomial series, the conclusion of Lemma 11.4 can be easily extended to the groups $T_\zeta(\cdot)$, $\zeta \in C$.

11.6 Corollary. For S, V as in Lemma 11.4 and t in a neighborhood of 0,

$$T_\zeta(t) = S(t)(I+itV)^\zeta = (I-itV)^{-\zeta}S(t) \qquad (\zeta \in C).$$

Proof. By [11; Theorem 13.2.2] and its proof,

$$T_\zeta(t) = \sum_{n=0}^{\infty} S_n(t)(i\zeta)^n \tag{11.6.1}$$

where

$$S_0(t) = S(t) \quad \text{and} \quad S_n(t)x = \int_0^t S(t-s)VS_{n-1}(s)x\, ds$$

$(n = 1,2,\ldots; t \in R; x \in X)$.

The known estimate $||S(t)|| \leq Me^{\omega|t|}$ implies that

$$||S_n(t)|| \leq M(M||V||\ |t|)^n e^{\omega|t|}/n!,$$

so that $\limsup_{n\to\infty} ||S_n(t)||^{1/n} = 0$.

Hence, for $t \in R$ fixed, $T_\zeta(t)$ is an entire function of ζ (represented by (11.6.1)) and

$$||T_\zeta(t)|| \leq Me^{\omega|t|}\exp(M||V||\ |t|\ |\zeta|). \tag{11.6.2}$$

For $|t| < ||V||^{-1}$, $(I+itV)^\zeta$ is an entire function of ζ, with the binomial series expansion

$$(I+itV)^\zeta = \sum_{n=0}^{\infty} \binom{\zeta}{n}(itV)^n.$$

Fix $\frac{\pi}{2} < \nu < \pi$. Since, uniformly in n,

$$\left|\binom{\zeta}{n}\right| < Ce^{\nu|\zeta|} \tag{11.6.3}$$

(cf. [11; p.234]), it follows that for $|t| < ||V||^{-1}$

$$||(I+itV)^{\zeta}|| < C(1-|t|\ ||V||)^{-1}e^{\nu|\zeta|}. \tag{11.6.4}$$

Fix t, $|t| < \delta = \min(\pi/M, 1)\, ||V||^{-1}$, and set

$$F_t(\zeta) = T_{\zeta}(t) - S(t)(I+itV)^{\zeta}.$$

By (11.6.2) and (11.6.4), $F_t(\cdot)$ is an entire function of exponential type $< \pi$, and $F_t(k) = 0$ for $k = 1,2,\ldots$ (by Lemma 11.4). It follows from [11, Theorem 3.13.7] that $F_t(\zeta) = 0$ for all $\zeta \in C$. This proves the first relation of Corollary 11.6; the second follows from the first as in the proof of Lemma 11.4.

11.7 Remark. For t in a neighborhood of 0, the following cocyle relation is satisfied

$$T_{\zeta+\omega}(t) = T_{\zeta}(t)S(-t)T_{\omega}(t)$$

$(\zeta,\omega \in C)$. The relation is meaningful for all real t, but it is unclear whether or not it is valid over R.

11.8 Definition. Let $V(\cdot)$ be a regular semigroup in C^+ (cf. Definition 6.0). The Nörlund function $\gamma(\cdot)$ of $V(\cdot)$ is given by

$$\gamma(\xi) = \limsup_{|\eta|\to\infty} |\eta|^{-1}\log||V(\xi+i\eta)|| \qquad (\xi > 0).$$

Let (α_0,α_1) be the largest α-interval such that the equation

$$\gamma(\xi) = \pi/2\alpha$$

has a solution (necessarily unique!)

$$\xi = \xi_0(\alpha) > 0 \quad \text{when} \quad 0 \le \alpha_0 < \alpha < \alpha_1 \le \infty \quad \text{(cf. [11; p.235])}.$$

11.9 Theorem. Let $V(\cdot)$ be a regular semigroup in C^+, and suppose $\alpha_1 > 1$. Then

$$V(\zeta) = \sum_{n=0}^{\infty} \binom{\zeta}{n}(V-I)^n \qquad (\zeta \in C^+)$$

where $V = V(1)$, with convergence in C^+ if $\alpha_0 \ge 1$ and in $\mathrm{Re}\,\zeta > \xi_0(1)$ if $\alpha_0 < 1$.

Proof. This is the special case of Theorem 17.6.1 in [11] with $\alpha = 1$ (which is valid, since $\alpha_1 > 1$).

11.10 Standing hypothesis. The setting for the abstract theory presented in the remainder of this section will be as follows:

(i) $S(\cdot)$ is a strongly continuous group of operators. Its infinitesimal generator is denoted by iS (with domain $D(S)$).

(ii) $V(\cdot)$ is a regular semigroup on C^+ with $\alpha_1 > 1$. We set $V = V(1)$.

(iii) $VD(S) \subset D(S)$ and $[S,V] \subset V^2$.

Recall the notation $T_\zeta = S+\zeta V$ $(\zeta \in C)$, and $T_\zeta(\cdot)$ (for the group generated by iT_ζ).

11.11 Theorem. For all $\zeta \in \overline{C}^+$, $V(\zeta)D(S) \subset D(S)$ and

(1) $[S,V(\zeta)] \subset \zeta V(\zeta+1)$.

Equivalently

(2) $V(\zeta)T_\zeta \subset SV(\zeta)$

and

(3) $V(\zeta)S \subset T_{-\zeta}V(\zeta)$.

Proof. The equivalence of the relations (1),(2), and (3) is trivial. We shall prove (1).

Set

$$V_k(\zeta) = \sum_{n=0}^{k} \binom{\zeta}{n} (V-1)^n , \qquad \mathrm{Re}\zeta > 0;\ k = 0,1,2,\dots .$$

Then $\quad V_k(\zeta)D(S) \subset D(S) \qquad (\zeta \in C^+; k = 1,2,\dots)$

and by Theorem 11.9, as $k \to \infty$,

$$V_k(\zeta)x \to V(\zeta)x$$

for all $x \in X$ and $\mathrm{Re}\zeta > \xi_0(1)$.

A simple induction (based on 11.10(iii) shows that

$$[S,V^j]x = jV^{j+1}x \qquad (x \in D(S); j=0,1,2,\dots).$$

Hence, for $x \in D(S)$ and $n = 1,2,\dots,$

$$[S,(V-I)^n]x = nV^2(V-I)^{n-1}x.$$

Therefore

$$SV_k(\zeta)x = \sum_{n=0}^{k} \binom{\zeta}{n} S(V-I)^n x$$

$$= \sum_{n=0}^{k} \binom{\zeta}{n} (V-I)^n Sx + V^2 \sum_{n=1}^{k} \binom{\zeta}{n} n(V-I)^{n-1} x.$$

Since $\binom{\zeta}{n} n = \binom{\zeta-1}{n-1}\zeta$ for $n = 1,2,\ldots$, we obtain

$$SV_k(\zeta)x = V_k(\zeta)Sx + \zeta V^2 V_{k-1}(\zeta-1)x$$

for $x \in D(S)$ and $k = 1,2,\ldots$.

If $\mathrm{Re}(\zeta-1) > \xi_0(1)$ and $x \in D(S)$, it follows from Theorem 11.9 that as $k \to \infty$

$$SV_k(\zeta)x \to V(\zeta)Sx + \zeta V^2 V(\zeta-1)x$$

$$= V(\zeta)Sx + \zeta V(\zeta+1)x.$$

Since S is a closed operator, we conclude that $V(\zeta)x \in D(S)$ and

$$SV(\zeta)x = V(\zeta)Sx + \zeta V(\zeta+1)x \tag{11.11.1}$$

for all $x \in D(S)$ and $\mathrm{Re}\zeta > \xi_0(1)+1$.

In particular, the function $\zeta \to SV(\zeta)x$ is analytic for $\mathrm{Re}\zeta > \xi_0(1)+1$ and the right hand side of (11.11.1) provides for it an analytic continuation in the half-plane $\mathrm{Re}\zeta > 0$. Since S is closed, it follows easily that for $x \in D(S)$ and $\mathrm{Re}\zeta > \xi_0(1)+1$, $\frac{d^n}{d\zeta^n} V(\zeta)x \in D(S)$ and $\frac{d^n}{d\zeta^n}(SV(\zeta)x) = S\frac{d^n}{d\zeta^n} V(\zeta)x$ $(n = 1,2,\ldots)$.

Therefore the Taylor series partial sums for $V(\cdot)x$ about ζ belong to $D(S)$, and when S is applied to them, they converge to $SV(\cdot)x$ in a suitable disc. Hence (S closed!) $V(\zeta)x \in D(S)$ for ζ in that disc, and consequently, for all $\zeta \in C^+$ (where $SV(\cdot)x$ admits an analytic continuation, as mentioned before), and (11.11.1) is valid throughout C^+.

Since $V(\cdot)$ is regular, letting $\xi = \mathrm{Re}\zeta \to 0+$ in (11.11.1), we obtain for $x \in D(S)$

$$SV(\xi+i\eta)x \to V(i\eta)Sx+i\eta V(i\eta+1)x.$$

Since also $V(\xi+i\eta)x \to V(i\eta)x$ and S is closed, it follows that $V(i\eta)x \in D(S)$ and

$$SV(i\eta)x = V(i\eta)Sx+i\eta V(i\eta+1)x.$$

This completes the proof of Theorem 11.11.

11.12 Corollary. For all $\zeta,\omega \in C$ with $\mathrm{Re}\zeta \geq \mathrm{Re}\omega$,

$$V(\zeta-\omega)T_\zeta \subset T_\omega V(\zeta-\omega) .$$

Proof. By Theorem 11.11(2),

$$\begin{aligned} V(\zeta-\omega)T_\zeta &= V(\zeta-\omega)[T_{\zeta-\omega}+\omega V] \\ &\subset SV(\zeta-\omega)+\omega V\cdot V(\zeta-\omega) \\ &= T_\omega V(\zeta-\omega). \end{aligned}$$

11.13. Corollary. T_ζ and T_ω are similar if $\mathrm{Re}\zeta = \mathrm{Re}\omega$:

$$T_\zeta = V(i\eta)^{-1}T_\omega V(i\eta) \tag{11.13.1}$$

for $i\eta = \zeta-\omega = i\mathrm{Im}(\zeta-\omega)$.

Proof. By Corollary 11.12 with $\mathrm{Re}\zeta = \mathrm{Re}\omega$,

$$T_\zeta \subset V(i\eta)^{-1}T_\omega V(i\eta)$$

for $i\eta = \zeta-\omega = i\mathrm{Im}(\zeta-\omega)$. Changing roles of ζ and ω (so that η is replaced by $-\eta$), we obtain

$$T_\omega \subset V(i\eta)T_\zeta V(i\eta)^{-1}.$$

The last two relations taken together are equivalent to (11.13.1).

We proceed to prove a converse of Corollary 11.13. We start with an elementary observation about similarity of unbounded operators.

11.14 Lemma. Let S and T be the infinitesimal generators of strongly continuous semigroups $S(\cdot)$ and $T(\cdot)$ respectively. Then S and T are similar (that is $S = Q^{-1}TQ$ with $Q \in B(X)$ nonsingular) if and only if $S(t) = Q^{-1}T(t)Q$ for all $t \geq 0$.

Proof. Suppose $S = Q^{-1}TQ$, and let $V(t) = Q^{-1}T(t)Q$, $t \geq 0$. Then $V(\cdot)$ is a strongly continuous semigroup. We must show that its generator V coincides with S. We have for $t > 0$

$$t^{-1}(T(t)-I)Q = Qt^{-1}(V(t)-I) .$$

Therefore, if $x \in D(V)$, then $Qx \in D(T)$, hence $x \in Q^{-1}D(T) \subset D(S)$, and $TQx = QVx$. Thus $Sx = Q^{-1}TQx = Vx$, that is $V \subset S$.

On the other hand, if $x \in D(S)$, then $Qx \in D(T)$ (since $S = Q^{-1}TQ$), hence $x \in D(V)$, i.e., $D(S) \subset D(V)$. This proves that $V = S$, and so $S(\cdot) = V(\cdot)$ as wanted.

The routine proof of the converse is left to the reader.

We now obtain appropriate versions of Theorems 6.8 and 6.9.

Recall that the Standing Hypothesis 11.10 implies that the boundary group $\{V(i\eta);\eta \in R\}$ exists and satisfies a growth condition $||V(i\eta)|| \leq Ke^{\nu|\eta|}$, $\eta \in R$.

11.15 <u>Theorem.</u> There exists a constant $H > 0$ such that

$$||T_{\xi+i\eta}(t)|| \leq H||S(t)||(1+|t|\ ||V||)^{|\xi|}e^{2\nu|\eta|} \qquad (t,\xi,\eta \in R).$$

<u>Proof.</u> The theorem follows from Lemma 11.4 in the same manner as Theorem 6.8 follows from Corollary 5.20. We present the proof anyway in order to indicate the slight changes needed in some of the details.

By Lemma 11.14 and Corollary 11.13,

$$T_{\xi+i\eta}(t) = V(-i\eta)T_\xi(t)V(i\eta) \qquad (t,\xi,\eta \in R). \tag{11.15.1}$$

For $t \in R$ fixed, set

$$\Phi_{it}(\zeta) = e^{\nu\zeta^2}T_\zeta(t) \qquad (\zeta \in C).$$

Then

$$||\Phi_{it}(\xi+i\eta)|| \leq K^2||T_\xi(t)||e^{\nu(\xi^2-\eta^2+2|\eta|)}$$
$$\leq K^2||T_\xi(t)||e^{\nu(\xi^2+1)}.$$

In particular, $||\Phi_{it}(\xi+i\eta)||$ is bounded in each strip $k-1 \leq \xi \leq k$ $(k \in Z)$.

By Lemma 11.4,

$$||\Phi_{it}(k+i\eta)|| \leq K^2||S(t)||e^{\nu(k^2+1)}(1+|t|\ ||V||)^{|k|}$$

for all $\eta \in R$ and $k \in Z$.

Since $\Phi_{it}(\cdot)$ is entire (cf. proof of Corollary 11.6), the "three-lines theorem" (cf. [5; Theorem VI.10.3]) applied in the strip $k-1 \leq \xi \leq k$ shows that

$$||\Phi_{it}(\xi+i\eta)|| \leq K^2||S(t)||(1+|t|\ ||V||)^{|\xi|}e^{\nu(\xi^2+5/4)},$$

hence

$$||T_\xi(t)|| = e^{-\nu\xi^2}||\Phi_{it}(\xi)|| \leq K^2||S(t)||(1+|t|\ ||V||)^{|\xi|}e^{5\nu/4},$$

and the theorem follows from (11.15.1), taking $H = K^4e^{5\nu/4}$.

11.16 Theorem. Suppose $||S(t)|| = O(1)$ $(t \in R)$. Then there exists a strictly positive function $C(\cdot)$ on R such that

$$||T_{\xi+i\eta}(t)|| \geq C(\xi)(1+|t|\,||V||)^{|\xi|}e^{-2\nu|\eta|} \quad (t,\xi,\eta \in R).$$

Proof. The proof of Theorem 6.9 carries over word for word, provided we make the following minor changes. First, write $T_\zeta(t)$ and $S(t)$ instead of $\exp(itT_\zeta)$ and $\exp(itS)$ respectively. We then deduce (6.8.5) from (11.15.1). Next, rather than using (6.8.6) (which is not available in our case for unrestricted t), we apply Corollary 11.5 to obtain the identities

$$T_{\xi+1}(t) = T_\xi(t)(I+itV)$$

and

$$T_{\xi-1}(t) = (I-itV)T_\xi(t) \qquad (11.16.1)$$

for all $t,\xi \in R$.

The proof of Theorem 6.9 proceeds now unchanged to the conclusion

$$\lim_{k\to\infty}(1+|t_k|\,||V||)^{-1}||T_{\pm 1}(t_k)|| = 0.$$

Since $||S(t)|| = O(1)$, it then follows from Lemma 11.4 that $\lim_{k\to\infty}(1+|t_k|\,||V||)^{-1}||I\pm it_kV|| = 0$, and the final contradiction is obtained.

11.17 Theorem. Suppose $||S(t)|| = O(1)$. Then, for $\zeta,\omega \in C$, the operators T_ζ and T_ω are similar if $\mathrm{Re}\zeta = \mathrm{Re}\omega$ and only if $|\mathrm{Re}\zeta| = |\mathrm{Re}\omega|$.

Proof. See proof of theorem 9.1. The slightly stronger version 9.2 is also valid in the present situation. Theorems 11.15 and 11.16 are used instead of theorems 6.8 and 6.9 respectively.

11.18 Example.

Let $1 < p < \infty$ and $\varepsilon > 0$. By Theorem 8.3, $\{J_\varepsilon^\zeta;\zeta \in C^+\}$ is a regular semigroup in $L^p(0,\infty)$, and for any fixed $\nu \in (\frac{\pi}{2},\pi)$, there exists a constant $C = C_{\nu,p}$ (independent of ε) such that $||J_\varepsilon^{i\eta}|| \leq Ce^{\nu|\eta|}$ $(\eta \in R)$.

Let $\gamma(\cdot)$ be the Norlund function of $\{J_\varepsilon^\zeta;\zeta \in C^+\}$.

$$\gamma(\xi) = \limsup_{|\eta|\to\infty}|\eta|^{-1}\log||J_\varepsilon^{\xi+i\eta}|| \leq \lim_{|\eta|\to\infty}|\eta|^{-1}[\log||J_\varepsilon^\xi||+\log C+\nu|\eta|] = \nu.$$

Since this is true for any $\nu \in (\frac{\pi}{2},\pi)$, therefore

$$\gamma(\xi) \leq \pi/2 \qquad (\xi > 0).$$

If $\alpha \in (\alpha_0,\alpha_1)$ (cf. 11.8), then

$$\pi/2\alpha = \gamma(\xi_0(\alpha)) \leq \pi/2,$$

i.e., $\alpha \geq 1$, and therefore $\alpha_0 \geq 1$. Hence $\alpha_1 > 1$, that is $\{J_\varepsilon^{\zeta}\}$ satisfies the Standing Hypothesis 11.10(ii).

Next, let $S(t): f(x) \to e^{itx}f(x)$, $t \in R$. $S(\cdot)$ is a strongly continuous group of operators in $L^p(0,\infty)$, with generator iS where $S = M: f(x) \to xf(x)$ with domain

$$D(M) = \{f \in L^p(0,\infty);\ xf(x) \in L^p(0,\infty)\}.$$

Clearly $||S(t)|| = 1$.

Let $f \in D(M)$. For all $x > 0$,

$$x(J_\varepsilon f)(x) = \int_0^x e^{-\varepsilon(x-t)}tf(t)dt + \int_0^x e^{-\varepsilon(x-t)}(x-t)f(t)dt$$

$$= (J_\varepsilon Mf)(x) + (J_\varepsilon{}^2 f)(x) \in L^p(0,\infty).$$

Hence $J_\varepsilon f \in D(M)$ (that is, $J_\varepsilon D(M) \subset D(M)$), and $[M,J_\varepsilon] \subset J_\varepsilon{}^2$. Thus the Standing Hypothesis (as well as the additional condition $||S(t)|| = O(1)$) is satisfied by the pair of operators (M,J_ε) in $L^p(0,\infty)$, $1 < p < \infty$. The results of this section are therefore valid for the operators $T_{\zeta,\varepsilon} = M+\zeta J_\varepsilon$ $(\varepsilon > 0,\ \zeta \in C)$ acting in $L^p(0,\infty)$ $(1 < p < \infty)$. In particular, $T_{\zeta,\varepsilon}$ and $T_{\omega,\varepsilon}$ are similar if $\mathrm{Re}\zeta = \mathrm{Re}\omega$ and only if $|\mathrm{Re}\zeta| = |\mathrm{Re}\omega|$.

11.19 C^k-classification.

Since T_ζ is the infinitesimal generator of a group of operators, it will be convenient to use the group $T_\zeta(\cdot)$ in order to define the C^k-classification. Let us start with some general considerations.

Suppose $\{T(t);\ t \in R\}$ is a strongly continuous group of operators on a Banach space X. Let M_T be the Banach algebra of all complex Borel measures m on R such that

$$||m||_T = \int_R ||T(t)||d|m|(t) < \infty$$

with the usual operations with measures and the norm $||\cdot||_T$ ($|m|$ denotes the total variation measure for m).

For any complex Borel measure m, let $\tilde{m}$ stand for its (inverse) Fourier-Stieltjes transform

$$\tilde{m}(s) = \int_R e^{its}dm(t) \quad (s \in R).$$

The image A_T of M_T under the map $m \to \tilde{m}$ is a Banach algebra of bounded continuous complex functions on R, with the usual pointwise operations and the norm $||\tilde{m}||_T = ||m||_T$. The algebra A_T contains in particular the exponentials

$$f_t(s) = e^{its} \qquad (s,t \in R),$$

since $f_t = \tilde{m}_t$, where m_t denotes the delta measure at t, and

$$||f_t||_T = ||m_t||_T = ||T(t)|| \qquad (t \in R).$$

For $f \in A_T$, define the operator $\tau(f)$ by

$$\tau(f)x = \int_R T(t)xdm(t), \quad x \in X, \ f = \tilde{m} \in A_T.$$

One verifies easily that τ is a continuous representation of the algebra A_T on X such that

$$\tau(f_t) = T(t) \qquad (t \in R).$$

We call τ the <u>natural operational calculus</u> for $T(\cdot)$.

Set

$$C_0^\infty(K) = \{f \in C_0^\infty;\ \text{supp } f \subset K\}$$

for any subset K of R.

Recall that an operator-valued distribution U is a continuous linear map from C_0^∞ to $B(X)$. The continuity of U means that for each compact subset K of R, there exists a constant B and an integer $k \geq 0$, such that

$$||U(f)|| \leq B \sum_{j \leq k} \sup|f^{(j)}|.$$

for all $f \in C_0^\infty(K)$.

If k can be chosen independently of K, then U is said to be of (finite) order $\leq k$, and the smallest such k is called the order of U.

If $T(\cdot)$ is <u>temperate</u> (that is, $||T(t)|| = O(|t|^h)$ for some finite h), then $S \subset M_T$ topologically, and therefore $S \subset A_T$ topologically since the map $m \to \tilde{m}$ restricted to S is a linear homeomorphism of S onto itself. In particular, $C_0^\infty \subset A_T$ topologically and so $\tau|C_0^\infty$ is a well-defined operator-valued distribution.

Let iT be the infinitesimal generator of the group $T(\cdot)$. <u>We say that T is of class C^k if $T(\cdot)$ is temperate and the operator distribution $\tau|C_0^\infty$ is of order $\leq k$.</u> The terminology is motivated by the fact that for each compact $K \subset R$, $\tau|C_0^\infty$ extends uniquely to a continuous representation on X of the Banach algebra $C_0^k(K) = \{f \in C_0^k; \text{supp } f \subset K\}$, that is, to a $C_0^k(K)$-operational calculus.

In case $T \in B(X)$, so that $T(t) = e^{itT}$, and T is of class C^k with real spectrum (in the sense of Sections 3 and 4), then $T(\cdot)$ is clearly temperate. Let (α,β) be an interval containing $\sigma(T)$, and let τ' denote the $C^k[\alpha,\beta]$-operational calculus for T. We saw in the proof of Lemma 3.2 that for $f \in C_0^k((\alpha,\beta))$, $\tau'(f) = \int_R \hat{f}(t)e^{itT}dt$, where $\hat{f}(t) = \frac{1}{2\pi}\int_R e^{-its}f(s)ds$. The natural operational calculus τ is given by $\tau(f) = \int_R g(t)e^{itT}dt$, with $f(t) = \check{g}(t) = \int_R e^{its}g(s)ds$. By the Fourier inversion formula, we clearly have $\hat{f} = g$, so that $\tau' = \tau$ on $C_0^\infty((\alpha,\beta))$. Since τ' is certainly a distribution of order $\leq k$, it follows that T is of class C^k in the present sense. Conversely, if T is of class C^k in the present sense (and is bounded), then since $T(\cdot)$ is temperate, it follows that T has real spectrum and is of class C^n for a suitable n (cf. Lemma 3.2). With (α,β) as before and $f \in C_0^\infty((\alpha,\beta))$, we have again $\tau'(f) = \tau(f)$, where τ' denotes the $C^n[\alpha,\beta]$-o.c. for T. Since τ is a distribution of order $\leq k$, $||\tau'(f)|| \leq B \sum_{j\leq k} \sup_{[\alpha,\beta]} |f^{(j)}|$ for all such f, and therefore τ' extends uniquely as a $C^k[\alpha,\beta]$ operational calculus for T. Hence T is of class C^k in the "old sense".

We may prove the following version of Theorem 6.10 in the present setting (cf. Standing Hypothesis).

11.20 Theorem. Let n,k be non-negative integers. If S is of class C^n, then T_ζ is of class C^{n+k} for all ζ in the strip $|\mathrm{Re}\zeta| \leq k$.

Proof. Since $S(\cdot)$ is temperate, the same is true of $T_\zeta(\cdot)$ by Theorem 11.15, for all $\zeta \in C$. Let τ_ζ denote the natural operational calculus for T_ζ. As in the proof of Corollary 11.6, one verifies easily that for each $f \in C_0^\infty$, $\tau_\zeta(f)$ is an entire function of ζ. Let $g \in S$ be such that $\tilde{g} = f$. Then by Lemma 11.4,

$$\begin{aligned}\tau_k(f)x &= \int_R g(t)T_k(t)x\,dt \\ &= \int_R g(t)S(t)(I+itV)^k x\,dt \\ &= \sum_{j=0}^{k} \binom{k}{j} V^j \int_R (it)^j g(t)S(t)x\,dt \\ &= \sum_{j=0}^{k} \binom{k}{j} V^j \tau_0(f^{(j)})x \quad \text{for all } x \in X.\end{aligned}$$

Since S is of class C^n, it follows that for each compact subset K of R

$$\begin{aligned}||\tau_k(f)|| &\leq \sum_{j=0}^{k} \binom{k}{j} ||V||^j B \sum_{r\leq n} \sup|f^{(j+r)}| \\ &\leq B' \sum_{j\leq n+k} \sup|f^{(j)}| \quad \text{for all } f \in C_0^\infty(K), \qquad (11.20.1)\end{aligned}$$

that is, T_k is of class C^{n+k}.

A similar computation starting from identity $T_{-k}(t) = (I-itV)^k S(t)$ (cf. Lemma 11.4) shows that T_{-k} is also of class C^{n+k}, i.e., the estimate (11.20.1) is satisfied by τ_{-k} as well.

Consider next the entire function

$$F(\zeta) = e^{\nu\zeta^2}\tau_\zeta(f) = e^{\nu\zeta^2} V(-i\eta)\tau_\xi(f)V(i\eta)$$

$(\tau = \xi+i\eta)$ for $f \in C_0^\infty$ fixed (cf. (11.15.1)). As in the proof of Theorem 11.15, we have

$$||F(\zeta)|| \leq K^2 e^{\nu(\xi^2+1)} ||\tau_\xi(f)||.$$

In particular, $F(\cdot)$ is bounded in the strip $|\mathrm{Re}\xi| \leq k$. By the "three lines theorem" and the estimates (11.20.1) for $\tau_{\pm k}(f)$, we obtain

$$||F(\zeta)|| \leq \mathrm{Const} \sum_{j\leq n+k} \sup|f^{(j)}|, \quad f \in C_0^\infty(K).$$

Hence, for ζ fixed in the strip $|\mathrm{Re}\zeta| \leq k$,

$$||\tau_\zeta(f)|| \leq \mathrm{Const} \sum_{j \leq n+k} \sup|f^{(j)}|, \quad f \in C_0^\infty(K),$$

that is, T_ζ is of class C^{n+k}.

Theorem 11.20 is valid in particular for the operators $T_{\zeta,\varepsilon} = M+\zeta J_\varepsilon$ in $L^p(0,\infty)$ ($\varepsilon > 0$, $1 < p < \infty$). Since M is obviously of class C ($= C^0$), the operators $T_{\zeta,\varepsilon}$ are of class C^k for $|\mathrm{Re}\zeta| \leq k$ (cf. Section 11.18).

12. Similarity (continued)

We saw in Section 9.6 that the operators $M+\zeta J_\varepsilon^\alpha$ $(\varepsilon \geq 0,\ \zeta \in C)$ acting in $L^p(0,N)$ $(N < \infty)$ are similar to M for all $\zeta \in C$ if $\mathrm{Re}\,\alpha \geq 1$, $\alpha \neq 1$, a situation contrasting drastically with that of Corollary 9.3 for the case $\alpha = 1$. For $\varepsilon > 0$, an analogous result is valid in $L^p(0,\infty)$. As before, this will follow from an abstract result about "perturbations" $S + C$, where S is closed and C is bounded.

12.1 Lemma. Let $V,C \in B(X)$ be commuting operators, and let S be a closed operator with domain $D(S)$. Then the following statements are equivalent:

(a) $VD(S) \subset D(S)$ and $[S,V] \subset C$.

(b) $R(\lambda;V)D(S) \subset D(S)$ and $S,R(\lambda;V) \subset CR(\lambda;V)^2$, $\lambda \in \rho(V)$.

(c) $g(V)D(S) \subset D(S)$ and $[S,g(V)] \subset Cg'(V)$, $g \in H(\sigma(V))$.

Proof. (a) ⇒ (b). By induction, one induces from (a) the relations

$$[S,V^j] \subset jV^{j-1}C \qquad (j = 1,2,\ldots).$$

Therefore (c) is valid for every polynomial g. Fix $\lambda \in \rho(V)$, and let $g_\lambda(\zeta) = (\lambda-\zeta)^{-1}$; g_λ is holomorphic in a neighborhood Ω of $\sigma(V)$. We can choose a sequence of polynomials $\{h_n\}$ such that $h_n \to g_\lambda$ uniformly on every compact subset of Ω. Then also $h_n' \to g_\lambda' = (\lambda-\zeta)^{-2}$ uniformly on every compact subset of Ω, and consequently $h_n(V) \to g_\lambda(V) = R(\lambda;V)$ and $h_n'(V) \to g_\lambda'(V) = R(\lambda;V)^2$ in the uniform operator topology. For $x \in D(S)$, one has by property (c) for polynomials:

$$h_n(V)x \in D(S),\ h_n(V)x \to R(\lambda;V)x, \text{ and}$$

$$Sh_n(V)x = h_n(V)Sx + Ch_n'(V)x \to R(\lambda;V)Sx + CR(\lambda;V)^2x$$

as $n \to \infty$. Since S is closed, it follows that $R(\lambda;V)x \in D(S)$ and

$$SR(\lambda;V)x = R(\lambda;V)Sx + CR(\lambda;V)^2x.$$

(b) ⇒ (c). Let $g \in H(\sigma(V))$, with analyticity domain Ω_g. Using the notation of Section 1, we choose $\Gamma = \Gamma(\Omega_g,\sigma(V))$, so that

$$g(V)x = \frac{1}{2\pi i}\int_\Gamma g(\lambda)R(\lambda;V)x\,d\lambda\ ,\quad x \in X.$$

Let $x \in D(S)$, and for $n = 1,2,\ldots,$ let x_n be the Riemann sum

$$x_n = \frac{1}{2\pi i} \sum_{j=1}^{n} (\lambda_j{}^n - \lambda_{j-1}{}^n) g(\lambda_j{}^n) R(\lambda_j{}^n; V) x$$

where $\lambda_j{}^n \in \Gamma$, $\lambda_0{}^n = \lambda_n{}^n$, and $\max\limits_{1 \leq j \leq n} |\lambda_j{}^n - \lambda_{j-1}{}^n| \to 0$ as $n \to \infty$. By (b), $x_n \in D(S)$, $x_n \to g(V)x$, and Sx_n are Riemann sums for the integral

$$\frac{1}{2\pi i} \int_\Gamma g(\lambda)[R(\lambda;V)Sx + CR(\lambda;V)^2 x]d\lambda = g(V)Sx + Cg'(V)x,$$

and therefore Sx_n converge to the last expression.

Since S is closed, it follows that $g(V)x \in D(S)$ and $Sg(V)x = g(V)Sx + Cg'(V)x$.

(c) $\Rightarrow$ (a). Take $g(\lambda) \equiv \lambda$ in (c).

12.2 Corollary. Let V,C be commuting bounded operators, and let S be a closed operator such that

$$VD(S) \subset D(S) \quad \text{and} \quad [S,V] \subset C.$$

Then for all $\zeta \in \mathbb{C}$, $S+\zeta C$ is similar to S with $e^{\zeta V}$ intertwining:

$$S+\zeta C = e^{-\zeta V} S e^{\zeta V}. \tag{12.2.1}$$

Proof. Fix $\zeta \in \mathbb{C}$, and take $g(\lambda) = e^{\pm\zeta\lambda}$ in Lemma 12.1 (c). Thus, for all $x \in D(S) = D(S+\zeta C)$,

$$Se^{\pm\zeta V}x = e^{\pm\zeta V}(S \pm \zeta C)x,$$

that is, $S+\zeta C \subset e^{-\zeta V} S e^{\zeta V}$

and $S \subset e^{\zeta V}(S+\zeta C)e^{-\zeta V}$.

The last two relations taken together are equivalent to (12.2.1).

By Lemma 12.1(c), we may replace V by $g(V)$ in Corollary 12.2, with $g \in H(\sigma(V))$. We then have

12.3 Corollary. Under the hypothesis of Corollary 12.2, for any $g \in H(\sigma(V))$, $S + g'(V)C$ is similar to S, with $e^{g(V)}$ intertwining:

$$S + g'(V)C = e^{-g(V)} S e^{g(V)}.$$

It is also easy to obtain the following "unbounded" versions of Corollary 9.8 and Corollary 9.9.

12.4 Corollary. Let $V_i, C_i \in B(X)$ commute with V_j $(1 \le i,j \le k)$. Suppose S is a closed operator with V_i-invariant domain $D(S) \subset X$, such that $[S,V_i] \subset C_i$ $(1 \le i \le k)$. Then for any choice of $g_i \in H(\sigma(V_i))$ $(1 \le i \le k)$,

$S + \sum_{i=1}^{k} g_i'(V_i)C_i$ is similar to S, with $\exp(\sum_{i=1}^{k} g_i(V_i))$ intertwining.

Proof. Set $S_0 = S$, and for a fixed choice of g_i

$$S_j = S + \sum_{i=1}^{j} g_i'(V_i)C_i \qquad (1 \le j \le k). \qquad (12.4.1)$$

Clearly, S_j is closed with V_{j+1}-invariant domain $D(S_j) = D(S)$ (for $j < k$), and

$$[S_j, V_{j+1}] = [S, V_{j+1}] \subset C_{j+1}.$$

Therefore, by Corollary 12.3,

$$S_{j+1} = S_j + g_{j+1}'(V_{j+1})C_{j+1} = \exp(-g_{j+1}(V_{j+1}))S_j \exp(g_{j+1}(V_{j+1}))$$

for $0 \le j < k$, and our corollary follows by recursion.

12.5 Corollary. Let $V_i, C_i \in B(X)$ commute with V_j $(1 \le i,j < \infty)$. Suppose S is a closed operator such that, for $i = 1,2,\ldots$

$$V_i D(S) \subset D(S) \quad \text{and} \quad [S,V_i] \subset C_i.$$

Let $g_i \in H(\sigma(V_i))$ $(i = 1,2,\ldots)$ be such that

$P = \sum_{i=1}^{\infty} g_i(V_i)$ and $S_\infty = S + \sum_{i=1}^{\infty} g_i'(V_i)C_i$ converge in the strong operator topology. Then S_∞ is similar to S with e^P intertwining: $S_\infty = e^{-P}Se^P$.

Proof. Define S_j by (12.4.1) and set $P_j = \sum_{i=1}^{j} g_i(V_i)$, where g_i are as in the corollary's statement. By the Uniform Boundedness Theorem, $K = \sup_j ||P_j|| < \infty$ since $P_j \to P$ strongly, and $||P|| \le K$. Therefore, as $j \to \infty$,

$$||(e^{P_j} - e^P)x|| = ||e^P(e^{P_j-P} - I)x|| \le (2K)^{-1}e^K(e^{2K}-1)||(P_j-P)x|| \to 0,$$

(where we used the commutativity assumption) that is, $e^{P_j} \to e^P$ strongly.

Now for $x \in D(S) = D(S_j)$ $(1 \le j \le \infty)$, we have

$$||e^{P_j}S_jx - e^PS_\infty x|| \le ||(e^{P_j} - e^P)S_\infty x|| + ||e^{P_j}(S_j - S_\infty)x||$$

$$\le ||(e^{P_j} - e^P)S_\infty x|| + e^K||(S_j - S_\infty)x|| \to 0 \quad \text{as } j \to \infty.$$

Let $x \in D(S)$ and $j = 1,2,\ldots$. By Corollary 12.4, $e^{P_j}x \in D(S)$, $e^{P_j}x \to e^Px$, and $Se^{P_j}x = e^{P_j}S_jx \to e^PS_\infty x$ by the preceding observation. Since S is closed, it follows that $e^Px \in D(S)$ and

$$Se^Px = e^PS_\infty x, \qquad x \in D(S) = D(S_\infty),$$

that is, $S_\infty \subset e^{-P}Se^P$. (12.5.1)

Clearly S_∞ satisfies the hypothesis of the Corollary on S. Replacing S by S_∞ and g_i by $-g_i$, S_∞ is then replaced by S and P by $-P$. Hence

$$S \subset e^PS_\infty e^{-P}.$$

Together with (12.5.1), this implies the wanted result.

Taking for example $g_i(\lambda) = \zeta_i\lambda$, where $\{\zeta_i\}$ is a complex sequence such that $\sum_{i=1}^{\infty} \zeta_iV_i$ and $\sum_{i=1}^{\infty} \zeta_iC_i$ converge strongly, we obtain that $S + \sum_{i=1}^{\infty} \zeta_iC_i$ is similar to S, with $\exp(\sum_{i=1}^{\infty} \zeta_iV_i)$ intertwining.

Consider now the situation of the Standing Hypothesis (cf. Section 11.10). Let $\alpha_i,\zeta_i \in C$ ($\mathrm{Re}\,\alpha_i \ge 1$, $\alpha_i \ne 1$) be such that $\sum_{i=1}^{\infty} \zeta_iV(\alpha_i)$ and $P = \sum_{i=1}^{\infty} \frac{\zeta_i}{\alpha_i-1} V(\alpha_i-1)$ converge strongly. Take $V_i = \frac{1}{\alpha_i-1} V(\alpha_i-1)$ in the preceding example. By Theorem 11.11, $V_iD(S) \subset D(S)$ and $[S,V_i] \subset V(\alpha_i) = C_i$. By the semigroup property of $V(\cdot)$, V_i and C_i commute with V_j $(i,j = 1,2,\ldots)$. Consequently, $S_\infty = S + \sum_{i=1}^{\infty} \zeta_iV(\alpha_i)$ is similar to S (with e^P intertwining). Clearly, the Standing Hypothesis 11.10 is satisfied when S is replaced by S_∞. Setting $T_\zeta = S_\infty+\zeta V$ $(\zeta \in C)$, we obtain the following generalization of Corollary 11.13, Theorem 11.17, Corollary 9.11, and Theorem 11.20.

12.6 <u>Theorem</u>. Let $S,V(\cdot)$ satisfy the Standing Hypothesis 11.10. For $i = 1,2,\ldots$, let $\alpha_i,\zeta_i \in C$ ($\mathrm{Re}\,\alpha_i \ge 1$, $\alpha_i \ne 1$) be such that $\sum_{i=1}^{\infty} \zeta_iV(\alpha_i)$ and $P = \sum_{i=1}^{\infty} \frac{\zeta_i}{\alpha_i-1} V(\alpha_i-1)$ converge strongly. Set $T_\zeta = S + \zeta V + \sum_{i=1}^{\infty} \zeta_iV(\alpha_i)$ $(\zeta \in C)$.

Then: (a) T_ζ and T_ω are similar if $\mathrm{Re}\zeta = \mathrm{Re}\omega$. In that case, $V(i\eta)$ intertwines T_ζ and T_ω, with $\eta = \mathrm{Im}(\zeta-\omega)$.

(b) If $S(\cdot)$ is a bounded group, then T_ζ and T_ω are similar if $\mathrm{Re}\zeta = \mathrm{Re}\omega$ and only if $|\mathrm{Re}\zeta| = |\mathrm{Re}\omega|$.

(c) T_ζ is similar to S if $\mathrm{Re}\zeta = 0$, with $V(i\eta)e^P$ intertwining ($\zeta = i\eta$). If $S(\cdot)$ is a bounded group, T_ζ is similar to S if and only if $\mathrm{Re}\zeta = 0$.

(d) Let n,k be non-negative integers. If S is of class C^n, then T_ζ is of class C^{n+k} for all ζ in the strip $|\mathrm{Re}\zeta| \leq k$.

Part (d) follows from the fact that if S is of class C^n (cf. Section 11.19), then the same is true of any operator similar to S (in particular, S_∞ is of class C^n).

Theorem 12.6 is valid in particular when $S = M$ and $V(\alpha) = J_\varepsilon{}^\alpha$ ($\alpha \in C^+$), for any $\varepsilon > 0$, and $X = L^p(0,\infty)$ with $1 < p < \infty$. Actually, since $J_{\varepsilon_1}{}^\alpha$ and $J_{\varepsilon_2}{}^\beta$ commute for any $\varepsilon_1, \varepsilon_2 > 0$ and $\alpha, \beta \in \overline{C}^+$, the result is true for the operators

$$T_\zeta = M + \zeta J_\varepsilon + \sum_{i=1}^{\infty} \zeta_i J_{\varepsilon_i}{}^{\alpha_i},$$

with any choice of $\varepsilon, \varepsilon_i > 0$, $\alpha_i \in C$ such that $\mathrm{Re}\alpha_i \geq 1$, $\alpha_i \neq 1$, and $\zeta_i \in C$, provided that the series $\sum_{i=1}^{\infty} \zeta_i J_{\varepsilon_i}{}^{\alpha_i}$ and $\sum_{i=1}^{\infty} \frac{\zeta_i}{\alpha_i - 1} J_{\varepsilon_i}{}^{\alpha_i - 1}$ converge strongly (cf. Section 11.18).

We may use Corollary 12.2 to obtain results analogous to Corollary 9.15 for $\Delta = (0,\infty)$ (or any unbounded interval on R), and to Corollary 9.16 for D an unbounded domain in C. We state the vector-valued versions below.

Let X be a Banach space. We use the notation $L^p(\Delta, X)$ of Section 9.15 for Δ an unbounded interval on R. Given a weight w on Δ and $h: \Delta \to B(X)$ such that $wh \in L^1(\Delta, B(X))$, we define the "weighted convolution operator" C_h on $L^p(\Delta, X)$ as before. Then $C_h \in B(L^p(\Delta, X))$. Let $M: f(x) \to xf(x)$ in $L^p(\Delta, X)$, with "maximal domain" $D(M) = \{f \in L^p(\Delta, X);\ Mf \in L^p(\Delta, X)\}$ (we use the notation M also for the "formal" multiplication operator, without specification of domain). Set $g(x) = h(x)/x$, and assume $\{wg \in L^1(\Delta, B(X)\}$. For $f \in D(M)$, we have

$$(MC_g f)(x) = \int_\Delta \frac{w(x)}{w(t)} xg(x-t)f(t)dt$$

$$= \int_\Delta \frac{w(x)}{w(t)} h(x-t)f(t)dt + \int_\Delta \frac{w(x)}{w(t)} g(x-t)tf(t)dt$$

$$= (C_h f)(x) + (C_g Mf)(x) \in L^p(\Delta,X)$$

since f and Mf are in $L^p(\Delta,X)$, that is, $C_g f \in D(M)$ and $[M,C_g] \subset C_h$. If the range of h in $B(X)$ is a commuting set of operators, it is clear that C_h commutes with C_g. By Corollary 12.2, we have

12.7 <u>Theorem</u>. Let w be a weight on the unbounded interval $\Delta \subset R$. Let $h: \Delta \to B(X)$ be such that $wh, wg \in L^1(\Delta, B(X))$, where $g(x) = h(x)/x$, and $h(x)$ commutes with $h(y)$ for all $x,y \in \Delta$. Then $M + C_h$ is similar to M, with $\exp(C_g)$ intertwining.

With notation as in Section 9.16, we have the following version of Corollary 9.16 for an unbounded domain $D \subset C$.

12.8 <u>Theorem</u>. Let $1 \leq p < \infty$, and let D be an unbounded domain in C. Let $h: D \to B(X)$ be measurable with commuting range, such that $wh \in (L^1 \cap L^\infty)(D,B(X))$. Then $M+C_h$ is similar to M, with $\exp(C_g)$ intertwining, where $g(z) = \frac{h(z)}{z}$.

<u>Proof</u>. As in the proof of Corollary 9.16, set

$$A(z,\zeta) = \frac{w(z)}{w(\zeta)} h(z-\zeta), \quad z,\zeta \in D.$$

The hypothesis $wh \in (L^1 \cap L^\infty)(D,B(X))$ implies condition (63) in [5; Theorem XX.2.10] for our kernel $A(z,\zeta)$. Therefore C_h and C_g are bounded operators (cf. [5], proof of Theorem XX.2.10). One verifies as in the preceding proof that $D(M)$ is C_g-invariant and $[M,C_g] \subset C_h$. Since C_g and C_h commute (because range(h) is a commuting set in $B(X)$), the conclusion of the theorem follows from Corollary 12.2.

A discrete analog of Theorem 12.7 (formulated in the scalar case for simplicity) goes as follows. Let $w = \{w_j\}_{j=1}^\infty$ be a "weight", that is a positive sequence such that $w_{n+m} \leq w_n w_m$ $(m,n = 1,2,\ldots)$. The weighted convolution operator $C_h = C_{h,w}$ on sequences $x = \{x_j\}_{j=1}^\infty$ is defined by

$$(C_h x)_k = \sum_{j=1}^{k-1} (w_k/w_j) h_{k-j} x_j \qquad k = 2,3,\ldots$$

$$= 0 \qquad k = 1,$$

where $h = \{h_j\}_{j=1}^{\infty}$ is any complex sequence. Since $w_k/w_j \leq w_{k-j}$ $(j < k)$, the operator C_h is bounded in ℓ^p $(1 \leq p \leq \infty)$, with norm not exceeding

$||wh||_1 = \sum_{j=1}^{\infty} w_j |h_j|$, whenever $||wh||_1 < \infty$. Let M denote the multiplication operator on sequences, $(Mx)_k = kx_k$ $(k = 1,2,\ldots)$.

Considered as an operator in ℓ^p with domain

$$D(M) = \{x \in \ell^p;\ Mx \in \ell^p\},$$

M is a closed operator.

Given h as above, let $g = \{g_j\}_{j=1}^{\infty}$ with $g_j = h_j/j$. Then $||wg||_1 \leq ||wh||_1 < \infty$, so that C_g is a bounded operator, which clearly commutes with C_h. If $x \in D(M)$, then for $k \geq 2$ (and trivially for $k = 1$)

$$(MC_g x)_k = \sum_{j=1}^{k-1} (w_k/w_j) k g_{k-j} x_j$$

$$= \sum_{j=1}^{k-1} (w_k/w_j) h_{k-j} x_j + \sum_{j=1}^{k-1} (w_k/w_j) g_{k-j} j x_j$$

$$= (C_h x)_k + (C_g Mx)_k.$$

Since both x and Mx are in ℓ^p, it follows that $MC_g x \in \ell^p$, that is $C_g D(M) \subset D(M)$, and $[M,C_g] \subset C_h$. By Corollary 12.2, we conclude that $M + C_h$ is similar to M, with $Q = \exp(C_g)$ intertwining: $M + C_h = Q^{-1}MQ$. If P is any non-singular operator in $B(\ell^p)$ intertwining $M + C_h$ and M, and $A = PQ^{-1}$, then A commutes with M (that is $AM \subset MA$). For $j = 1,2,\ldots$, denote by e^j the sequence with $(e^j)_k = \delta_{jk}$. Clearly $e^j \in D(M)$ and $Me^j = je^j$. Therefore $([M,A]e^j)_k = (k-j)(Ae^j)_k = 0$ $(k,j = 1,2,\ldots)$. It follows that $(Ae^j)_k = \delta_{jk}\lambda_j$, where $\lambda_j = (Ae^j)_j$. Hence A is the diagonal operator (or multiplication operator) $(Ax)_k = \lambda_k x_k$ $(k = 1,2,\ldots)$. Necessarily, the sequence $\lambda = \{\lambda_k\}_{k=1}^{\infty}$ is bounded and bounded away from zero (since A is bounded and non-singular). We formalize the preceding observations:

12.9 Proposition. For any weight $w = \{w_j\}_{j=1}^{\infty}$ and any complex sequence $h = \{h_j\}_{j=1}^{\infty}$ such that $||wh||_1 < \infty$, $M + C_h$ is similar to M, and the most general intertwining operator P for $M + C_h$ and M (that is $M + C_h = Q^{-1}MQ$) is of the form $P = A\exp(C_g)$, where A is a multiplication operator $(Ax)_j = \lambda_j x_j$ with $\{\lambda_j\}$ and $\{1/\lambda_j\}$ bounded, and $g_j = h_j/j$ $(j = 1,2,\ldots)$.

If for example $w \in \ell^1$; we may take $h_j = \zeta$ $(j = 1,2,\ldots)$ with $\zeta \in C$ fixed. Thus $M+\zeta N_w$ is similar to M for all $\zeta \in C$, where N_w denotes the "weighted summation operator" in ℓ^p:

$$(N_w x)_k = \sum_{j=1}^{k-1} (w_k/w_j)x_j \qquad (k \geq 2)$$

$$(N_w x)_1 = 0.$$

When $w_j = e^{-\varepsilon j}$ for $\varepsilon > 0$, N_w is the discrete analog of J_ε acting in $L^p(0,\infty)$. However, in contrast to the present situation, we saw in Section 11.18 that $M+\zeta J_\varepsilon$ (acting in $L^p(0,\infty)$) is similar to M if and only if $\operatorname{Re}\zeta = 0$.

By Proposition 12.9, the most general intertwining operator for $M+\zeta N_w$ and M is of the form $P = A\exp(\zeta H_w)$, where A is a non-singular bounded multiplication operator $((Ax)_k = \lambda_k x_k)$ and H_w is the weighted convolution operator $C_g = C_{g,w}$ with $g_j = 1/j$, that is

$$(H_w x)_k = \sum_{j=1}^{k-1} \frac{w_k}{w_j}(k-j)^{-1}x_j \qquad \text{for } k > 1$$

and $(H_w x)_1 = 0$.

We observe the following explicit form of the intertwining holomorphic group $\exp(\zeta H_w)$:

$$[\exp(\zeta H_w)x]_k = \sum_{j=1}^{k} \frac{w_k}{w_j}(-1)^{k-j}\binom{-\zeta}{k-j}x_j \qquad (k = 1,2,\ldots)$$

(where $\binom{\alpha}{0} = 1$ as usual, for all $\alpha \in C$).

To see this, denote by $G_w(\zeta)$ the operator defined by the righthand side. It is a weighted convolution operator $C_{h(\zeta)}$ plus the identity, where $h(\zeta)_j = (-1)^j\binom{-\zeta}{j}$. Since $h(\zeta) \in \ell^\infty$ (for each $\zeta \in C$; cf. (11.6.3)) and $w \in \ell^1$, $G_w(\zeta) \in B(\ell^p)$.

We verify the group property for $G_w(\cdot)$. Fix $x \in \ell^p$, and $\alpha,\beta \in C$. For $n = 1,2,\ldots,$

$$[G_w(\alpha)G_w(\beta)x]_n = \sum_{k=1}^{n}\sum_{j=1}^{k}\frac{w_n}{w_j}(-1)^{n-j}\binom{-\alpha}{n-k}\binom{-\beta}{k-j}x_j$$

$$= \sum_{j=1}^{n}\frac{w_n}{w_j}\sum_{r=0}^{n-j}\binom{-\alpha}{n-j-r}\binom{-\beta}{r}(-1)^{n-j}x_j$$

$$= \sum_{j=1}^{n}\frac{w_n}{w_j}(-1)^{n-j}\binom{-\alpha-\beta}{n-j}x_j = [G_w(\alpha+\beta)x]_n.$$

Clearly, $\{G_w(t);t \geq 0\}$ is strongly continuous and $G_w(0) = I$. Actually $G_w(\cdot)$ is holomorphic in C.

A simple calculation shows that the infinitesimal generator of $G_w(\cdot)$ is equal to H_w, and consequently $G_w(\zeta) = \exp(\zeta H_w)$ as claimed above.

Note that $G_w(1) = I + N_w$, so that the operator $T_\zeta = M+\zeta G_w(1)$ is similar to $M+\zeta I$. In particular, for $\zeta,\omega \in C$, T_ζ is similar to T_ω if and only if $\zeta = \omega$. This should be compared to the similarity result of Section 11.13 for the continuous analog $T_\zeta = M+\zeta J_\varepsilon^{\ 1}$ in $L^p(0,\infty)$: T_ζ is similar to T_ω if $\mathrm{Re}\zeta = \mathrm{Re}\omega$ and only if $|\mathrm{Re}\zeta| = |\mathrm{Re}\omega|$. On the other hand, we saw in Section 12.6 that $M+\zeta J_{\varepsilon_1}^{\ \alpha}$ is similar to $M+\omega J_{\varepsilon_2}^{\ \beta}$ for all $\zeta,\omega \in C$, for any choice of $\varepsilon_1,\varepsilon_2 > 0$ and α,β in the half plane $\mathrm{Re} z \geq 1$, $z \neq 1$.

An analogous, but "weaker" phenomenon is exhibited by the following example. Consider the multiplication operator $M: f(x) \to xf(x)$ with maximal domain in $L^p(R)$: $D(M) = \{f \in L^p(R);\ Mf \in L^p(R)\}$. Let $D = d/dx$ with domain

$$\{f \in L^p(R);\ \ f \text{ absolutely continuous and } f' \in L^p(R)\}.$$

Then $A = \frac{1}{2}D^2$ is the infinitesimal generator of the Gauss-Wierstrass semigroup $\{T(t);\ t > 0\}$, where

$$[T(t)f](x) = (2\pi t)^{-\frac{1}{2}} \int_R e^{-u^2/2t} f(x-u)\,du \qquad (x \in R).$$

Consider the "perturbations" $M-tD$ $(t > 0)$ with domain equal to the intersection of the domains of M and D. If $f \in \mathrm{domain}(M-tD)$, then for $x \in R$, and integration by parts gives

$$[MT(t)f](x) = (2\pi t)^{-\frac{1}{2}} \int_R e^{-u^2/2t}[(x-u)+u]f(x-u)\,du$$

$$= [T(t)Mf](x) - t[T(t)Df](x) \in L^p(R),$$

that is, $T(t)f \in D(M)$ and

$$T(t)(M-tD) \subset MT(t) \qquad (t > 0).$$

This is a "half-similarity" relation between M and $M-tD$, with the semigroup $T(t)$ intertwining.

Looking for common ground for the various examples encountered above, we note that in all cases, intertwining is achieved by a semigroup (or group) (e.g., $e^{\zeta V}$ in Corollary 12.2; $V(i\eta)$ in Theorem 12.6; $T(t)$ in the last example). If A denotes the infinitesimal generator of that semigroup, the intertwined operators are $S-tC$ (or $S-\zeta C$) and S, where $[A,S] \subset C$ and C commutes with A (in the sense $[A,C] \subset 0$, where 0 denotes the zero operator). Denoting by d_A the derivation induced by A acting on arbitrary operators U (that is, $d_A U = [A,U]$ with the appropriate domain), the above relations between A, S and C take the form

$$d_A S \subset C$$

$$d_A{}^2 S \subset 0.$$

We shall consider below the problem of recuperating the similarity results in such an abstract setting.

12.10 Lemma. Let A,S be linear operators acting in a Banach space X, with domains $D(A)$ and $D(S)$ respectively. Then for all $\lambda \in \rho(A)$,

$$[R_\lambda, S] \supset R_\lambda (d_A S) R_\lambda,$$

where $R_\lambda = R(\lambda;A)$.

Proof. Fix $\lambda \in \rho(A)$, and let x belong to the domain of $R_\lambda(d_A S) R_\lambda$, that is,

$$R_\lambda x \in D(d_A S). \tag{12.10.1}$$

Equivalently, $R_\lambda x \in D(S)$, $AR_\lambda x \in D(S)$, and $SR_\lambda x \in D(A)$. Hence $x = \lambda R_\lambda x - AR_\lambda x \in D(S)$, and so

$$x \in D([R_\lambda, S]) = \{x \in D(S); R_\lambda x \in D(S)\}.$$

By (12.10.1), since $d_A S = [S, \lambda I - A]$, we obtain

$$R_\lambda (d_A S) R_\lambda x = R_\lambda \{S(\lambda I - A) - (\lambda I - A)S\} R_\lambda x = [R_\lambda, S]x.$$

12.11 Lemma. Suppose $d_A{}^2 S \subset 0$ and let $\lambda \in \rho(A)$.
Then if $x \in X$ is such that

(H_n) $\qquad R_\lambda{}^k x \in D(d_A{}^2 S) \qquad (k = 1,\ldots,n),$

one has the relations

$$(R_n)\quad [R_\lambda^{\;k},S]x = kR_\lambda^{\;k+1}(d_AS)x$$
$$[R_\lambda^{\;k},d_AS]x = 0 \qquad (k = 1,\ldots,n).$$

Proof. We proceed by induction on n. By Lemma 12.10, $[R_\lambda,d_AS] \supset R_\lambda(d_A^{\;2}S)R_\lambda$. Therefore, if $x \in X$ is such that $R_\lambda x \in D(d_A^{\;2}S)$, then one has

$$x \in D(d_AS);\ R_\lambda x \in D(d_AS)$$

and $$[R_\lambda,d_AS]x = 0.$$

By Lemma 12.10, it follows that $[R_\lambda,S]x = R_\lambda^{\;2}(d_AS)x$. Hence (H_1) implies (R_1). Now, assume (H_n) implies (R_n) for some n, and suppose $x \in X$ satisfies (H_{n+1}). Since $D(d_A^{\;2}S) \subset D(d_AS)$, we have $R_\lambda^{\;k}x \in D(d_AS)$ for $k = 1,\ldots,n+1$. In particular, it follows as in the proof of Lemma 12.10 that $R_\lambda^{\;k}x \in D(S)$ for $k = 0,\ldots,n+1$. Moreover

$$[R_\lambda^{\;n+1},S]x = R_\lambda[R_\lambda^{\;n},S]x + [R_\lambda,S]R_\lambda^{\;n}x.$$

Since $R_\lambda^{\;n}x$ satisfies (H_1) and x satisfies (H_n), the first part of the proof and the induction hypothesis imply that

$$[R_\lambda^{\;n+1},S]x = R_\lambda nR_\lambda^{\;n+1}(d_AS)x + R_\lambda^{\;2}(d_AS)R_\lambda^{\;n}x$$
$$= (n+1)R_\lambda^{\;n+2}(d_AS)x,$$

and

$$R_\lambda^{\;n+1}(d_AS)x = R_\lambda[(d_AS)R_\lambda^{\;n}]x = [R_\lambda(d_AS)]R_\lambda^{\;n}x$$
$$= (d_AS)R_\lambda^{\;n+1}x.$$

Thus (H_{n+1}) implies (R_{n+1}), and the lemma is proved.

We now assume that S is a closed operator, and A is the infinitesimal generator of a strongly continuous semigroup $\{T(t);t > 0\}$ of class (C_0). Let $\omega_0 = \lim_{t\to\infty} t^{-1}\log||T(t)||$, and fix a (real) sequence $\Lambda = \{\lambda_n\}$ such that $\lambda_n > \omega_0$ for all n and $\lambda_n \to \infty$. In particular, $\Lambda \subset \rho(A)$. Suppose $\mathcal{D}$ is a linear manifold contained in $D(d_A^{\;2}S)$ and R_λ-invariant for all $\lambda \in \Lambda$. For $x \in \mathcal{D}$, Condition (H_n) is satisfied for all n and all $\lambda \in \Lambda$; hence the relations (R_n) are valid for all n and $\lambda \in \Lambda$, provided $d_A^{\;2}S \subset 0$. Let C be any closed extension of $d_A S$. Since

$$\mathcal{D} \subset D(d_A{}^2 S) \subset D(d_A S) \subset D(A) \cap D(S) \cap D(C),$$

the restrictions of A, S and C to $\mathcal{D}$ are well-defined; we denote them by A_0, S_0 and C_0 respectively.

Finally, we assume that $T(t)$ is a one-to-one operator for some (hence for all) $t > 0$.

12.12 Lemma. $S_0 - tC_0$ is closable and

$$T(t)\overline{(S_0 - tC_0)} \subset \overline{S_0}\, T(t) \qquad (t > 0)$$

In addition, $T(t)\overline{C_0} \subset \overline{C_0} T(t) \qquad (t > 0).$

Proof. For $\lambda \in \rho(A)$ and $n = 0,1,\ldots,$ set

$$H_{\lambda,n} = \sum_{k=0}^{n} (t\lambda^2 R_\lambda)^k / k!$$

and $H_\lambda = \exp(t\lambda^2 R_\lambda) \qquad (t > 0)$.

Fix $x \in \mathcal{D}$. By Lemma 12.11,

$$[H_{\lambda,n}, S_0]x = \sum_{k=1}^{n} [(t\lambda^2)^k / (k-1)!] R_\lambda^{k+1} C_0 x$$

$$= t(\lambda R_\lambda)^2 H_{\lambda,n-1} C_0 x \qquad (12.12.1)$$

for all $\lambda \in \Lambda$ and $n = 1,2,\ldots$.

Now $H_{\lambda,n} x \in \mathcal{D} = D(S_0)$, $H_{\lambda,n} x \to H_\lambda x$ as $n \to \infty$, and by (12.12.1),

$$S_0 H_{\lambda,n} x = H_{\lambda,n} S_0 x - t(\lambda R_\lambda)^2 H_{\lambda,n-1} C_0 x$$

$$\to H_\lambda S_0 x - t(\lambda R_\lambda)^2 H_\lambda C_0 x \qquad \text{as } n \to \infty.$$

Since S_0 is closable (because S is closed), it follows that $H_\lambda \mathcal{D} \subset D(\overline{S_0})$, and

$$\overline{S_0} H_\lambda x = H_\lambda S_0 x - t(\lambda R_\lambda)^2 H_\lambda C_0 x \qquad (12.12.2)$$

for all $\lambda \in \Lambda$ and $x \in \mathcal{D}$.

Next, $e^{-\lambda t} H_\lambda x \in D(\overline{S_0})$ for all $\lambda \in \Lambda$, $x \in \mathcal{D}$, and $t > 0$, and $e^{-\lambda t} H_\lambda x \to T(t)x$ as $\lambda \to \infty$ through Λ (the latter fact is valid for all $x \in X$, by formula (11.7.2), p. 352 in [11]). By (12.12.2), since $\lambda R_\lambda \to I$ in the strong operator topology as $\lambda \to \infty$, we have

$$\overline{S}_0 e^{-\lambda t} H_\lambda x \to T(t)S_0 x - tT(t)C_0 x$$

as $\lambda \to \infty$ through Λ (for $t > 0$ and $x \in \mathcal{D}$).

Since $\overline{S}_0$ is closed, it follows that

$$T(t)\mathcal{D} \subset D(\overline{S}_0) \tag{12.12.3}$$

and

$$\overline{S}_0 T(t)x = T(t)S_0 x - tT(t)C_0 x \tag{12.12.4}$$

for all $x \in \mathcal{D}$ and $t > 0$.

Equivalently

$$T(t)(S_0 - tC_0) \subset \overline{S}_0 T(t) \quad (t > 0). \tag{12.12.5}$$

Since $\overline{S}_0 T(t)$ is closed, $T(t)(S_0-tC_0)$ is closable for each $t > 0$. Suppose $\{x_n\} \subset \mathcal{D}$ converges to zero and $(S_0-tC_0)x_n \to y$ as $n \to \infty$ (for some $t > 0$). Then $T(t)(S_0-tC_0)x_n \to T(t)y$, and therefore $T(t)y = 0$. Since $T(t)$ is one-to-one, we conclude that $y = 0$, that is, S_0-tC_0 is closable. Now, by (12.12.5),

$$T(t)\overline{(S_0-tC_0)} \subset \overline{T(t)(S_0-tC_0)} \subset \overline{S}_0 T(t) \qquad (t > 0).$$

By Lemma 12.11,

$$[H_{\lambda,n}, C_0]x = 0 \quad \text{for all} \quad x \in \mathcal{D},\ \lambda \in \Lambda, \quad \text{and} \quad n = 0,1,\dots .$$

Starting with this relation instead of (12.12.1), the argument yielding from (12.12.1) to (12.12.3) and (12.12.4) shows that $T(t)\mathcal{D} \subset D(\overline{C}_0)$ and $\overline{C}_0 T(t) = T(t)C_0 x$ for all $x \in \mathcal{D}$ and $t > 0$. Thus $T(t)C_0 \subset \overline{C}_0 T(t)$, and since the right hand side is closed, we obtain

$$T(t)\overline{C}_0 \subset \overline{T(t)C_0} \subset \overline{C}_0 T(t) \qquad (t > 0).$$

<u>12.13.</u> The setting for the following theorem will be as follows: We are given <u>closed</u> operators S and C; A is the infinitesimal generator of a strongly continuous <u>group</u> of operators $\{T(t); t \in R\}$. Let $\omega_0 = \max(\omega_0^+, \omega_0^-)$, where $\omega_0^{\pm} = \lim_{t\to\infty} t^{-1}\log\|T(\pm t)\|$, and fix a symmetric real sequence Λ such that $|\lambda| > \omega_0$ and $|\lambda| \to \infty$ on Λ. Suppose $\mathcal{D}$ is a linear manifold contained in $D(d_A{}^2 S)$ and R_λ-invariant for all $\lambda \in \Lambda$.

We assume the commutation relations

$$d_A S \subset C$$

and $\qquad d_A{}^2 S \subset 0.$

Recall the following definition [39, p.190]: Let P,Q be operators acting in a Banach space X. We say that P is Q-bounded if $D(P) \supset D(Q)$ and there exist non-negative constants a and b such that

$$||Px|| \le a||x|| + b||Qx||$$

for all $x \in D(Q)$.

When P and Q are closable and P is Q-bounded, one sees easily that $\overline{P}$ is $\overline{Q}$-bounded; in particular, $D(\overline{P}) \supset D(\overline{Q})$.

We now apply Lemma 12.12 to the semigroups $\{T^{\pm}(t); t > 0\}$, where $T^{\pm}(t) = T(\pm t)$ $(t > 0)$. Thus $S_0 - tC_0$ is closable for all $t \in R$ and

$$\overline{S_0 - tC_0} \subset T(-t)\overline{S}_0 T(t) \qquad (t \in R). \tag{12.13.1}$$

For all $t \in R$ and $x \in \mathcal{D}$, we have (again by Lemma 12.12)

$$\begin{aligned} T(t)S_0 x &= \overline{S}_0 T(t)x + tT(t)C_0 x \\ &= (\overline{S}_0 + t\overline{C}_0)T(t)x. \end{aligned}$$

Applying $T(-t)$ to both sides and replacing t by $-t$ we obtain

$$S_0 \subset T(t)(\overline{S}_0 - t\overline{C}_0)T(-t) \qquad (t \in R). \tag{12.13.2}$$

Suppose now that C_0 is S_0-bounded, that is

$$||C_0 x|| \le a||x|| + b||S_0 x|| \qquad (x \in \mathcal{D}) \tag{12.13.3}$$

for some non-negative constants a and b.

If $x \in D(\overline{S}_0 - t\overline{C}_0) = D(\overline{S}_0)$, there exists a sequence $\{x_n\} \subset \mathcal{D}$ such that $x_n \to x$ and $\{S_0 x_n\}$ is a Cauchy sequence. By (12.13.3), $\{C_0 x_n\}$ is a Cauchy sequence, and it follows that $x \in D(\overline{S_0 - tC_0})$ and

$$\overline{S}_0 - t\overline{C}_0 \subset \overline{S_0 - tC_0} \qquad (t \in R). \tag{12.13.4}$$

The same argument applies when S_0 is C_0-bounded.

Consequently, by (12.13.2), we have for all $t \in R$,

$$\overline{S}_0 \subset T(t)(\overline{S_0 - tC_0})T(-t) \;. \tag{12.13.5}$$

Together with (12.13.1), this last relation implies that

$$\overline{S_0-tC_0} = T(-t)\overline{S}_0T(t) \tag{12.13.6}$$

for all $t \in R$. We formalize these conclusions in the following result.

12.14 Theorem. (Setting 12.13). For all $t \in R$, S_0-tC_0 is closable, and the following relations are valid

$$\overline{S_0-tC_0} \subset T(-t)\overline{S}_0T(t)$$

and $$S_0 \subset T(t)(\overline{S}_0-t\overline{C}_0)T(-t).$$

In case C_0 is S_0-bounded or vice versa, $\overline{S_0-tC_0}$ is similar to $\overline{S}_0$, with $T(t)$ intertwining:

$$\overline{S_0-tC_0} = T(-t)\overline{S}_0T(t) \qquad (t \in R).$$

The similarity result is true in particular when C_0 (or S_0) is bounded. In this case, one has $\overline{S_0-tC_0} = \overline{S}_0-t\overline{C}_0$ $(t \in R)$. If $\mathcal{D}$ is a core for the closed operators S and C, that is, $\overline{S}_0 = S$ and $\overline{C}_0 = C$, then $\overline{S_0-tC_0} = S-tC$, i.e., $S-tC$ is closed, $\mathcal{D}$ is a core for it, and the similarity result takes the form

$$S-tC = T(-t)ST(t) \qquad (t \in R).$$

Formally:

12.16 Corollary. (Setting 12.13) Suppose $\mathcal{D}$ is a core for S and C, and C_0 (or S_0) is bounded. Then, for all $t \in R$, $S-tC$ is closed, $\mathcal{D}$ is a core for it, and $S-tC$ is similar to S, with $T(t)$ intertwining.

We consider now the case when A generates a regular semigroup $\{T(\zeta);\zeta \in C^+\}$ (cf. Definition 6.0). As before, S and C are closed operators such that $d_AS \subset C$ and $d_A{}^2S \subset 0$; $\Lambda \subset \rho(A)$ is a real sequence tending to ∞, and one fixes a linear manifold $\mathcal{D} \subset D(d_A{}^2S)$ which is R_λ-invariant for all $\lambda \in \Lambda$. In this setting, we shall prove the following:

12.16 Theorem. (1) For all $\zeta \in \overline{C^+}$, $S_0-\zeta C_0$ is closable, and the following relations are valid:

$$T(\zeta)(\overline{S_0-\zeta C_0}) \subset \overline{S}_0T(\zeta) \tag{12.16.1}$$

$$T(\zeta)\overline{C}_0 \subset \overline{C}_0T(\zeta). \tag{12.16.2}$$

(2) If either C_0 is S_0-bounded or vice versa, then $\overline{S_0-i\eta C_0}$ is similar to $\overline{S}_0$:

$$\overline{S_0-i\eta C_0} = T(-i\eta)\overline{S}_0T(i\eta) \qquad (\eta \in R). \tag{12.16.3}$$

(3) If C_0 (or S_0) is bounded, then for all $\zeta,\omega \in \overline{C^+}$, $\overline{S_0-\zeta C_0}$ $(= \overline{S}_0-\zeta\overline{C}_0)$ is similar to $\overline{S_0-\omega C_0}$ $(= \overline{S}_0-\omega\overline{C}_0)$ whenever $\mathrm{Re}\zeta = \mathrm{Re}\omega$, with $T(i\mathrm{Im}(\zeta-\omega))$ intertwining. In particular, if $\mathcal{D}$ is a core for S and C, then $S-\zeta C$ is closed with core $\mathcal{D}$, and $S-\zeta C$ is similar to $S-\omega C$ whenever $\mathrm{Re}\zeta = \mathrm{Re}\omega$.

Proof. (1) By Theorem 6.1, for $\xi > 0$ and $\eta \in R$,

$$\log||T(\xi+i\eta)|| \le \log||T(\xi)|| + \log||T(i\eta)||$$
$$\le \log||T(\xi)|| + \log K + \nu|\eta|$$

where $0 \le \nu < \infty$. Therefore $\gamma(\xi) \le \nu$ (cf. Definition 11.8), and consequently $\pi/2\alpha = \gamma(\xi_0(\alpha)) \le \nu$, that is, $\alpha \ge \pi/2\nu$, for any $\alpha \in (\alpha_0,\alpha_1)$. Hence $\alpha_0 \ge \pi/2\nu > 0$ (since $\nu < \infty$).

Fix $\zeta \in C^+$ and ε such that $0 < \varepsilon < \min(\alpha_0,\mathrm{Re}\zeta)$. Since $\alpha_0 > 0$, we have the convergent binomial expansion (cf. [11, Theorem 17.6.1])

$$T(\zeta) = \sum_{n=0}^{\infty} \binom{\zeta/\varepsilon}{n}(T(\varepsilon)-I)^n.$$

Set

$$T_k(\zeta) = \sum_{n=0}^{k} \binom{\zeta/\varepsilon}{n}(T(\varepsilon)-I)^n \qquad (k = 0,1,\ldots).$$

By (12.12.3),

$$T_k(\zeta)\mathcal{D} \subset D(\overline{S}_0) \qquad (k = 0,1,\ldots).$$

For $x \in \mathcal{D}$, we have by (12.12.4), for $k = 1,2,\ldots$,

$$\overline{S}_0T_k(\zeta)x = \sum_{n=0}^{k}\binom{\zeta/\varepsilon}{n}\sum_{j=0}^{n}\binom{n}{j}(-1)^{n-j}\overline{S}_0T(j\varepsilon)x$$

$$= \sum_{n=0}^{k}\binom{\zeta/\varepsilon}{n}\sum_{j=0}^{n}\binom{n}{j}(-1)^{n-j}T(\varepsilon)^jS_0x - \sum_{n=1}^{k}\binom{\zeta/\varepsilon}{n}\sum_{j=1}^{n}\binom{n}{j}(-1)^{n-j}j\varepsilon T(\varepsilon)^jC_0x$$

$$= T_k(\zeta)S_0x - \sum_{n=1}^{k}\binom{\zeta/\varepsilon}{n}n\varepsilon(T(\varepsilon)-I)^{n-1}T(\varepsilon)C_0x$$

$$= T_k(\zeta)S_0x - \sum_{n=1}^{k}\binom{\zeta/\varepsilon-1}{n-1}\zeta(T(\varepsilon)-I)^{n-1}T(\varepsilon)C_0x$$

$$= T_k(\zeta)S_0x - \zeta T_{k-1}(\zeta-\varepsilon)T(\varepsilon)C_0x \to T(\zeta)S_0x - \zeta T(\zeta)C_0x \quad \text{as } k \to \infty,$$

since $\mathrm{Re}(\zeta-\varepsilon) > 0$.

It follows that $T(\zeta)\mathcal{D} \subset D(\overline{S}_0)$, and

$$T(\zeta)(S_0-\zeta C_0) \subset \overline{S}_0 T(\zeta) \qquad (\zeta \in C^+). \tag{12.16.4}$$

By Theorem 6.2, $T(\zeta)$ is one-to-one for each $\zeta \in C^+$. It then follows from (12.16.4), by the argument made after (12.12.5), that $S_0-\zeta C_0$ is closable and (12.16.1) is valid in C^+. Next, for $x \in \mathcal{D}$ and $\xi+i\eta \in C^+$, we have $T(\xi+i\eta)x \in D(\overline{S}_0)$, $T(\xi+i\eta)x \to T(i\eta)x$ as $\xi \to 0$, and by (12.16.4),

$$\overline{S}_0 T(\xi+i\eta)x = T(\xi+i\eta)(S_0-(\xi+i\eta)C_0)x \to T(i\eta)(S_0-i\eta C_0)x \quad \text{as} \quad \xi \to 0.$$

Since $\overline{S}_0$ is closed, it follows that $T(i\eta)\mathcal{D} \subset D(\overline{S}_0)$, and $T(i\eta)(S_0-i\eta C_0) \subset \overline{S}_0 T(i\eta)$. As before, this implies that $S_0-i\eta C_0$ is closable and (12.16.1) is valid throughout $\overline{C^+}$. The relation (12.16.2) is proved by a similar (but simpler) calculation, using the second identity of Lemma 12.12.

(2) If $x \in \mathcal{D}$, we have by (12.16.4) with $\zeta = -i\eta$:

$$\begin{aligned} T(-i\eta)S_0 x &= \overline{S}_0 T(-i\eta)x - i\eta T(-i\eta)C_0 x \\ &= (\overline{S}_0 - i\eta\overline{C}_0)T(-i\eta)x \end{aligned}$$

(cf. (12.16.2)). Hence, for all $\eta \in R$,

$$S_0 \subset T(i\eta)(\overline{S}_0-i\eta\overline{C}_0)T(-i\eta). \tag{12.16.5}$$

If either C_0 is S_0-bounded or vice versa, we have $\overline{S}_0-i\eta\overline{C}_0 \subset \overline{S_0-i\eta C_0}$ (cf. (12.13.4)), and (12.16.5) implies that $\overline{S}_0 \subset T(i\eta)(\overline{S_0-i\eta C_0})T(-i\eta)$. Together with (12.16.1) for $\zeta = i\eta$, this relation proves (12.16.3).

(3) If C_0 (or S_0) is bounded, $S_0-\zeta C_0$ is closable for all $\zeta \in C$, and $\overline{S_0-\zeta C_0} = \overline{S}_0-\zeta\overline{C}_0$. Suppose $\zeta,\omega \in \overline{C^+}$ have equal real parts, and let $i\eta = \zeta-\omega$. Then, by (12.16.3) and (12.16.2),

$$\begin{aligned} \overline{S_0-\zeta C_0} &= \overline{S}_0-\zeta\overline{C}_0 = (\overline{S}_0-i\eta\overline{C}_0)-\omega\overline{C}_0 \\ &= T(-i\eta)\overline{S}_0 T(i\eta)-\omega\overline{C}_0 \\ &\subset T(-i\eta)(\overline{S}_0-\omega\overline{C}_0)T(i\eta) \\ &= T(-i\eta)(\overline{S_0-\omega C_0})T(i\eta). \end{aligned}$$

Interchanging ζ and ω, the needed reversed inclusion is obtained, and the proof is complete.

12.17 <u>Corollary</u>. Let S and C be closed operators in the Banach space X, and let $A \in B(X)$ be such that $AD(S) \subset D(S)$, $d_A S \subset C$, and $d_A{}^2 S \subset 0$. Then $S-\zeta C$ is similar to S for all $\zeta \in \mathbb{C}$, with $e^{\zeta A}$ intertwining.

<u>Proof</u>. One verifies easily that

$$D(d_A{}^2 S) = D(d_A S) = D(S) \subset D(C).$$

By induction, one sees that

$$SA^k x = A^k Sx + kA^{k-1}Cx \quad (k = 1,2,\dots;x \in D(S)).$$

For $|\lambda| > ||A||$, let $x_n = \sum_{k=0}^{n} \lambda^{-k-1} A^k x \quad (n = 1,2,\dots;x \in D(S))$.

Then $x_n \in D(S)$, $x_n \to R_\lambda x$, and

$$Sx_n = \sum_{k=0}^{n} \lambda^{-k-1} A^k Sx + \sum_{k=0}^{n} \lambda^{-k-1} kA^{k-1} Cx \to R_\lambda Sx - R_\lambda{}^2 Cx \quad \text{as } n \to \infty.$$

Since S is closed, it follows that $R_\lambda D(S) \subset D(S)$ when $|\lambda| > ||A||$.

Let Λ be any (real) sequence such that $\lambda > ||A||$ and $\lambda \to \infty$ on Λ. Since $D(S) = D(d_A{}^2 S)$ is R_λ-invariant for all $\lambda \in \Lambda$, we may apply Theorem 12.16 with $\mathcal{D} = D(S)$, and with the (regular) semigroups $T^{\pm}(\zeta) = e^{\pm\zeta A}$ $(\zeta \in \mathbb{C}^+)$ (correspondingly, we take $\pm C$ instead of C). We then obtain $S-\zeta C \subset e^{-\zeta A} S e^{\zeta A}$ for all $\zeta \in \mathbb{C}$. In particular, $e^{\zeta A} D(S) \subset D(S)$ $(\zeta \in \mathbb{C})$, and therefore the domain of $e^{-\zeta A} S e^{\zeta A}$ is precisely $D(S)$, which proves the wanted similarity relation.

Of course, the above corollary may be proved directly in a straightforward manner.

Finally, given $\Lambda \subset \rho(A)$, note that the maximal linear manifold $\mathcal{D}$ with the required properties is

$$\mathcal{D} = \mathcal{D}(\Lambda) = \{x \in D(d_A{}^2 S); R_\lambda{}^k x \in D(d_A{}^2 S); k=1,2,\dots,\lambda \in \Lambda\}.$$

We conclude our discussion of similarity with a result corresponding to the sufficiency part of Corollary 9.3 <u>for $L^p(0,\infty)$</u>, $1 < p < \infty$. (Cf. also Example 11.18; the case considered presently is when $\varepsilon = 0$.) The operators

$$M: f(x) \to xf(x)$$

and

$$J: f(x) \to \int_0^x f(t)\,dt,$$

with maximal domains in $L^p(0,\infty)$

$$D(M) = \{f \in L^p(0,\infty); Mf \in L^p(0,\infty)\}$$

and $$D(J) = \{f \in L^p(0,\infty); Jf \in L^p(0,\infty)\},$$

are closed densely defined operators. The preceding general theory is too weak to give the wanted result. Although an abstract theory is available (cf. [15]), we shall content ourselves with a direct proof of the following

12.18 Theorem. Let $1 < p < \infty$ and $\zeta,\omega \in C\smallsetminus\{0\}$. Then, considered as operators with maximal domains in $L^p(0,\infty)$, $M+\zeta J$ and $M+\omega J$ are similar whenever $\mathrm{Re}\zeta = \mathrm{Re}\omega$.

The proof will be achieved through a series of lemmas.

Let J^ζ denote the Riemann-Liouville operator

$$(J^\zeta f)(x) = \frac{1}{\Gamma(\zeta)}\int_0^x (x-t)^{\zeta-1} f(t)\,dt$$

with maximal domain in $L^p(0,\infty)$

$$D(J^\zeta) = \{f \in L^p(0,\infty); J^\zeta f \in L^p(0,\infty)\}.$$

It can be shown that J^ζ is closed and densely defined (cf. [14; Proposition 4.1]), but we shall not use this fact.

For $\varepsilon > 0$, we consider the regular semigroups $\{J_\varepsilon^\zeta; \mathrm{Re}\zeta > 0\}$ (cf. Definition 8.1 and Theorem 8.3(2)), and their boundary groups $\{J_\varepsilon^{i\eta}; \eta \in R\}$ (cf. Theorem 6.1 and Theorem 8.3(4)). We first concern ourselves with the behavior of $J_\varepsilon^{i\eta}$ as $\varepsilon \to 0+$.

12.19 Lemma. As $\varepsilon \to 0+$, $J_\varepsilon^{i\eta}$ converges to a bounded operator (denoted $J^{i\eta}$) in the weak operator topology on $B(L^p(0,\infty))$, for each $\eta \in R$. The family $\{J^{i\eta}; \eta \in R\}$ is a strongly continuous group of operators on $L^p(0,\infty)$, and

$$J^{\xi+i\eta} = J^\xi J^{i\eta} = J^{i\eta} J^\xi \qquad (\xi > 0, \eta \in R)$$

on $D(J^{\xi+i\eta}) = D(J^\xi)$.

For any $\nu \in (\frac{\pi}{2},\pi)$, there exists a positive constant $C = C_{\nu,p}$ such that $||J^{i\eta}|| \le Ce^{\nu|\eta|}$.

Proof. Fix $\eta \in R$ and $f \in L^p(0,\infty)$. By Theorem 8.3(4), $||J_\varepsilon^{i\eta} f||_p$ is bounded with respect to $\varepsilon > 0$. Since $L^p(0,\infty)$ is reflexive for $1 < p < \infty$, the set

$\{J_\varepsilon^{i\eta}f; \varepsilon > 0\}$ has weak limit points in $L^p(0,\infty)$. Let g be such a point. For $\xi > 0$, the boundary group $J_\varepsilon^{i\eta}$ satisfies

$$(J_\varepsilon^{\xi} J_\varepsilon^{i\eta} f)(x) = (J_\varepsilon^{\xi+i\eta} f)(x) \qquad (x > 0),$$

and since $0 \le 1-e^{\varepsilon(t-x)} \le \varepsilon(x-t)$ for $0 \le t \le x$, we have

$$|J_\varepsilon^{\xi} \cdot J_\varepsilon^{i\eta} f(x) - J^{\xi+i\eta} f(x)| \le |\Gamma(\xi+i\eta)|^{-1} \int_0^x (1-e^{\varepsilon(t-x)})(x-t)^{\xi-1} |f(t)|dt$$

$$\le |\Gamma(\xi+i\eta)|^{-1} \varepsilon \int_0^x (x-t)^{\xi} |f(t)|dt,$$

that is, as $\varepsilon \to 0+$,

$$J_\varepsilon^{\xi} J_\varepsilon^{i\eta} f(x) \to J^{\xi+i\eta} f(x) \qquad (\xi, x > 0). \tag{12.19.1}$$

On the other hand,

$$J_\varepsilon^{\xi} J_\varepsilon^{i\eta} f(x) = \Gamma(\xi)^{-1} \int_0^x (x-t)^{\xi-1} J_\varepsilon^{i\eta} f(t)dt + R_\varepsilon, \tag{12.19.2}$$

where, by Hölder's inequality,

$$|R_\varepsilon| \le \varepsilon\Gamma(\xi)^{-1} \int_0^x (x-t)^{\xi} |J_\varepsilon^{i\eta} f(t)|dt$$

$$\le \varepsilon\Gamma(\xi)^{-1} (\xi p'+1)^{-1/p'} x^{\xi+1/p'} ||J_\varepsilon^{i\eta} f|| = O(\varepsilon)$$

by Theorem 8.3(4).

Letting $\varepsilon \to 0$ through a sequence such that $J_\varepsilon^{i\eta} f \to g$ weakly, the first term on the right of (12.19.2) converges to $J^\xi g(x)$ for each fixed $x > 0$, provided $\xi > 1/p$ (this ensures that $(x-t)^{\xi-1}$ is in $L^{p'}(0,x)$). In light of (12.19.1), we conclude that

$$J^\xi g = J^{\xi+i\eta} f \qquad (\eta \in R, \xi > 1/p).$$

For $\xi > 0$ arbitrary, $1+\xi > 1 > 1/p$, and therefore

$$J(J^\xi g) = J^{1+\xi} g = J^{1+\xi+i\eta} f = J(J^{\xi+i\eta} f).$$

Since J is one-to-one on the locally integrable functions, it follows that

$$J^\xi g = J^{\xi+i\eta} f \qquad (\xi > 0, \eta \in R). \tag{12.19.3}$$

Suppose $h \in L^p(0,\infty)$ is also a weak limit point of $\{J_\varepsilon^{i\eta}; \varepsilon > 0\}$. Taking $\xi = 1$ in (12.19.3) and in the corresponding relation for h, we obtain $Jh = Jg$, and therefore $h = g$ as elements of $L^p(0,\infty)$. This proves the <u>existence</u> of the weak limit of $J_\varepsilon^{i\eta}f$ as $\varepsilon \to 0+$. Denote it by $J^{i\eta}f$. By (12.19.3),

$$J^{\xi}J^{i\eta}f = J^{\xi+i\eta}f \qquad (\xi > 0, \eta \in R), \tag{12.19.4}$$

for all $f \in L^p(0,\infty)$. Also, by Theorem 8.3(4),

$$||J^{i\eta}|| \leq Ce^{\nu|\eta|} \qquad (\eta \in R),$$

for any $\nu \in (\frac{\pi}{2},\pi)$, with C depending only on ν and p.

Suppose $f \in D(J^\xi)$ for some $\xi > 0$. Then $J^{i\eta}(J^\xi f) \in L^p(0,\infty)$, and by (12.19.4) and Fubini's theorem, $JJ^{i\eta}(J^\xi f) = J^{1+i\eta}(J^\xi f) = J^{1+\xi+i\eta}f = J(J^{\xi+i\eta}f)$. Hence $J^{i\eta}(J^\xi f) = J^{\xi+i\eta}f$ as elements of $L^p(0,\infty)$. In particular, $J^{\xi+i\eta}f \in L^p(0,\infty)$, that is

$$D(J^\xi) \subset D(J^{\xi+i\eta}).$$

The same argument with $J^{-i\eta}$ and $J^{\xi+i\eta}$ gives the reversed inclusion. Hence $D(J^{\xi+i\eta}) = D(J^\xi)$, and on this domain, $J^{\xi+i\eta} = J^{i\eta}J^\xi = J^\xi J^{i\eta}$ (cf. (12.19.4)).

The group property of $J^{i\eta}$ can be proved as follows. For each $f \in L^p(0,\infty)$ and $\eta, \delta \in R$, we have by (12.19.4)

$$\begin{aligned} J^2J^{i(\eta+\delta)}f &= J^{(1+i\eta)+(1+i\delta)}f = J^{1+i\eta}J^{1+i\delta}f \\ &= J^{1+i\eta}JJ^{i\delta}f = JJ^{1+i\eta}J^{i\delta}f = J^2J^{i\eta}J^{i\delta}f, \end{aligned}$$

and therefore $J^{i(\eta+\delta)} = J^{i\eta}J^{i\delta}$ since J^2 is one-to-one.

For each $f \in L^p(0,\infty)$, $J^{i\eta}f$ is a weakly measurable function of η with values in $L^p(0,\infty)$, as the weak limit of the strongly continuous functions $J_\varepsilon^{i\eta}f$ (as $\varepsilon \to 0+$). Since $L^p(0,\infty)$ is separable, $J^{i\eta}f$ is strongly measurable (cf. Corollary 2 of Theorem 3.5.3 in [11]). Due to the group property, it follows that $\{J^{i\eta}; \eta \in R\}$ is strongly continuous, by [11; Theorem 10.2.3]. This completes the proof of the lemma.

12.20 <u>Lemma</u>. Let $0 \leq n-1 < \xi \leq n$, where n is an integer. Then for each $f \in D(J^n)$, $J^{\xi+i\eta}f$ is the weak limit of $J_\varepsilon^{\xi+i\eta}f$ as $\varepsilon \to 0+$ $(\eta \in R)$.

<u>Proof</u>. We first verify by induction that for each locally integrable f, $\varepsilon > 0$, and $k = 1,2,\ldots,$

$$J_\varepsilon^k f = (1-\varepsilon J_\varepsilon)^k J^k f. \tag{12.20.1}$$

For $k = 1$, we have indeed

$$J_\varepsilon f(x) + \varepsilon J_\varepsilon Jf(x) = \int_0^x [e^{\varepsilon(t-x)}f(t) + \varepsilon e^{\varepsilon(t-x)}Jf(t)]dt$$

$$= \int_0^x \frac{d}{dt}[e^{\varepsilon(t-x)}(Jf)(t)]dt = Jf(x).$$

Suppose (12.20.1) is valid for some k. Then (using the case $k = 1$ as well),

$$\begin{aligned} J_\varepsilon^{k+1}f &= J_\varepsilon J_\varepsilon^k f = J_\varepsilon(1-\varepsilon J_\varepsilon)^k J^k f \\ &= (1-\varepsilon J_\varepsilon)^k J_\varepsilon J^k f = (1-\varepsilon J_\varepsilon)^k(1-\varepsilon J_\varepsilon)JJ^k f \\ &= (1-\varepsilon J_\varepsilon)^{k+1}J^{k+1}f, \quad \text{as wanted.} \end{aligned}$$

Now, if $f \in D(J^k)$ for some k $(k = 1,2,\ldots)$, then by (12.20.1), Theorem 8.3 and Theorem 6.1,

$$\begin{aligned} ||J_\varepsilon^{k+i\eta}f|| &= ||J_\varepsilon^{i\eta}J_\varepsilon^k f|| = ||J_\varepsilon^{i\eta}(1-\varepsilon J_\varepsilon)^k J^k f|| \\ &\leq ||J_\varepsilon^{i\eta}||\ ||1-\varepsilon J_\varepsilon||^k ||J^k f|| \\ &\leq Ce^{\nu|\eta|}(1+||\varepsilon J_\varepsilon||)^k ||J^k f|| \\ &\leq C2^k e^{\nu|\eta|}\ ||J^k f|| \qquad (\eta \in R), \end{aligned}$$

since, as a convolution operator, $\varepsilon J_\varepsilon$ has norm smaller than or equal to the $L^1(0,\infty)$-norm of $\varepsilon e^{-\varepsilon t}$, which is equal to 1. The last inequality is trivially true for $k = 0$ as well (cf. Theorem 8.3(4)). We now apply the three lines theorem to the function

$$\Phi(\zeta) = \exp(\nu\zeta^2)J_\varepsilon^{\zeta}f \qquad (\zeta = \xi+i\eta)$$

in the strip $n-1 \leq \xi \leq n$ (n integer), for $f \in D(J^n)(\subset D(J^{n-1}))$ fixed. We obtain from the preceding estimate

$$||J_\varepsilon^{\xi+i\eta}f|| \leq C2^\xi e^{\nu|\eta|}\ ||J^n f||^{\xi-n+1}||J^{n-1}f||^{n-\xi},$$

where C depends only on ν and p.

In particular, $||J_\varepsilon^{\xi+i\eta}f||$ is bounded with respect to $\varepsilon > 0$, for fixed $f \in D(J^n)$. As before, let g be a weak limit point of the set $\{J_\varepsilon^{\xi+i\eta}f;\varepsilon > 0\} \subset L^p(0,\infty)$ (ξ,η, and f fixed), say $g = \text{weak}\lim_{n\to\infty} J_{\varepsilon_n}^{\xi+i\eta}f$.

For each fixed $x > 0$, the characteristic function of the interval $[0,x]$ belongs to $L^{p'}(0,\infty)$, and therefore $Jg(x) = \lim_{n\to\infty}\int_0^x J_{\varepsilon_n}^{\xi+i\eta}f(t)dt.$

By (12.19.1), the integrand converges pointwise to $J^{\xi+i\eta}f(t)$, and a simple estimate shows that it is dominated by $\Gamma(\xi)|\Gamma(\xi+i\eta)|^{-1}J^{\xi}|f|(t) \in L^1(0,x)$ ($\xi, x > 0$ fixed). By dominated convergence, it follows that

$$Jg = JJ^{\xi+i\eta}f,$$

hence $g = J^{\xi+i\eta}f$ (as elements of $L^p(0,\infty)$), and the lemma follows.

12.21 Lemma. M is a "closed" operator with respect to weak sequential convergence in $L^p(0,\infty)$; that is, if $\{f_n\} \subset D(M)$, $f_n \overset{w}{\to} f$, and $Mf_n \overset{w}{\to} g$, then $f \in D(M)$ and $Mf = g$ ($\overset{w}{\to}$ denotes weak convergence).

Proof. For h in the domain of M in $L^{p'}(0,\infty)$,

$$\int_0^\infty hg\,dx = \lim_{n\to\infty} \int_0^\infty h(Mf_n)\,dx = \lim_{n\to\infty} \int_0^\infty (Mh)f_n\,dx$$

$$= \int_0^\infty (Mh)f\,dx = \int_0^\infty h(Mf)\,dx.$$

Since $D(M)$ (in $L^{p'}(0,\infty)$) is dense in $L^{p'}(0,\infty)$, it follows that $Mf \in L^p(0,\infty)$, that is, $f \in D(M)$, and $Mf = g$, as wanted.

Proof of Theorem 12.18. Let $\zeta,\omega \in \mathbb{C}\setminus\{0\}$ be such that $\mathrm{Re}\,\zeta = \mathrm{Re}\,\omega$, and let $\eta = \mathrm{Im}(\zeta-\omega)$. By Corollary 11.13 (cf. Example 11.18),

$$J_\varepsilon^{i\eta}(M+\zeta J_\varepsilon) = (M+\omega J_\varepsilon)J_\varepsilon^{i\eta} \qquad (\varepsilon > 0),$$

that is, for each $f \in D(M)$, $J_\varepsilon^{i\eta}f \in D(M)$ and

$$MJ_\varepsilon^{i\eta}f = J_\varepsilon^{i\eta}Mf + (\zeta-\omega)J_\varepsilon^{1+i\eta}f.$$

Suppose now that $f \in D(M) \cap D(J) = D(M+\zeta J) = D(M+\omega J)$ (since $\zeta,\omega \neq 0$). By Lemma 12.19 and Lemma 12.20 (with $n = 1$), we have

$$D(M) \ni J_\varepsilon^{i\eta}f \overset{w}{\to} J^{i\eta}f,$$

and

$$MJ_\varepsilon^{i\eta}f \overset{w}{\to} J^{i\eta}Mf + (\zeta-\omega)J^{1+i\eta}f.$$

Therefore, by Lemma 12.21, $J^{i\eta}f \in D(M)$ and

$$MJ^{i\eta}f = J^{i\eta}Mf + (\zeta-\omega)J^{1+i\eta}f$$

$$= J^{i\eta}Mf + \zeta J^{i\eta}Jf - \omega JJ^{i\eta}f,$$

where we used again Lemma 12.19. Hence

$$(M+\omega J)J^{i\eta}f = J^{i\eta}(M+\zeta J)f$$

for all $f \in D(M+\omega J) = D(M+\zeta J)$.

This completes the proof of Theorem 12.18, and establishes also that the boundary group $J^{i\eta}$ constructed in Lemma 12.19 implements the similarity.

13. Singular C^n-operators

Consider again our model example $T_\zeta = M+\zeta J$ acting in $L^p(0,1)$, $1 < p < \infty$, and let n be a positive integer. By Theorem 8.4 and Corollary 10.12, the set of complex numbers ζ for which T_ζ is of class C^n is precisely the strip $|\mathrm{Re}\,\zeta| \le n$, while the set of ζ for which T_ζ is spectral is the imaginary axis. This illustrates strikingly the rough statement that a C^n-operator is "rarely" spectral.

On the other hand, any spectral operator T of type n (with real spectrum), that is, with a quasinilpotent part N such that $N^{n+1} = 0$, is clearly of class C^n. This follows for example from Proposition 2.3.1, since T has the Jordan decomposition $T = S+N$ with S and N commuting, and with S spectral of scalar type, hence obviously of class $C(\sigma(T))$ (if E is the resolution of the identity for S, set $\tau(f) = \int_{\sigma(T)} f\,dE$ for $f \in C(\sigma(T))$; cf. [5; Part III]).

This brings us to the topic of this section: determine restricted classes of C^n-operators consisting only of spectral operators. We begin with the simple case $n = 0$.

13.1 <u>Theorem</u>. Let X be a weakly complete Banach space, and let $K \subset \mathbb{C}$ be compact. Then $T \in B(X)$ is of class $C(K)$ if and only if T is spectral of scalar type with spectrum in K.

<u>Proof</u>. (for the "only if" part). Let $T \in B(X)$ be of class $C(K)$. By Proposition 1.2, $\sigma(T) \subset K$. Denote by τ the $C(K)$-operational calculus for T. For each $x \in X$, the map $f \to \tau(f)x$ is a bounded linear operator from $C(K)$ to X. Since X is weakly complete, the map is weakly compact, by Theorem VI.7.6 in [5], and therefore, by [5; Theorem VI.7.3], there exists an X-valued measure m_x on $B(K)$ (= the σ-algebra of all Borel subsets of K), such that

(a) x^*m_x is a countably additive regular Borel measure for each $x^* \in X^*$;

(b) $\tau(f)x = \int_K f(\lambda)m_x(d\lambda) \qquad (f \in C(K))$;

and

(c) $||m_x(\delta)|| \le ||\tau||\ ||x|| \qquad (\delta \subset B(K))$.

The uniqueness of the representation (b) implies that, for each $\delta \in \mathcal{B}(K)$, the map

$$E(\delta): x \to m_x(\delta)$$

is linear, and $\|E(\delta)\| \leq \|\tau\|$ by (c).

By (a), $x^* E(\cdot)x$ is a countably additive regular Borel measure for each $x \in X$ and $x^* \in X^*$. It follows in particular that $E(\cdot)x$ is strongly countably additive.

According to (b),

$$\tau(f) = \int_K f(\lambda)E(d\lambda) \qquad (f \in C(K)).$$

In particular, $E(K) = \tau(1) = I$ and $T = \tau(\lambda) = \int_K \lambda E(d\lambda)$.

It remains to show that $E(\delta \cap \sigma) = E(\delta)E(\sigma)$ for all $\delta, \sigma \in \mathcal{B}(K)$.

For all $f, g \in C(K)$,

$$\int_K f(\lambda)g(\lambda)E(d\lambda) = \tau(fg) = \tau(f)\tau(g)$$
$$= \int_K f(\lambda)E(d\lambda)\tau(g).$$

By the uniqueness statement of the Riesz representation theorem, $g(\lambda)E(d\lambda) = E(d\lambda)\tau(g)$, that is, if χ_δ denotes the characteristic function of $\delta \in \mathcal{B}(K)$, then

$$\int_K g(\lambda)\chi_\delta(\lambda)E(d\lambda) = \int_\delta g(\lambda)E(d\lambda) = E(\delta)\tau(g)$$
$$= \int_K g(\lambda)E(\delta)E(d\lambda)$$

for all $g \in C(K)$ and $\delta \in \mathcal{B}(K)$.

Again by uniqueness, $\chi_\delta(\lambda)E(d\lambda) = E(\delta)E(d\lambda)$, that is,

$$E(\delta)E(\sigma) = \int_\sigma \chi_\delta(\lambda)E(d\lambda)$$
$$= \int_K \chi_{\delta\cap\sigma}(\lambda)E(d\lambda) = E(\delta\cap\sigma)$$

for all $\delta, \sigma \in \mathcal{B}(K)$, Q.E.D.

13.2 <u>Corollary</u>. Let H be a Hilbert space. Then $T \in B(H)$ is spectral of scalar type if and only if it is similar to a normal operator.

<u>Proof</u>. By Theorems 13.1 and 2.1.1, $T \in B(H)$ is spectral of scalar type if and only if

$$T \in [C(\sigma(T))]_{B(H)} = S_{B(H)}(\sigma(T)).$$

13.3 Corollary. Let H be a Hilbert space, and let $E: \mathcal{B}(K) \to B(H)$ be a spectral measure. Then there exists a non-singular operator $Q \in B(H)$ such that $QE(\cdot)Q^{-1}$ is a selfadjoint spectral measure.

Proof. The operator $T = \int_K \lambda E(d\lambda)$ is scalar type spectral with spectrum in K. By Corollary 13.2, there exists a non-singular operator $Q \in B(H)$ such that $N = QTQ^{-1}$ is normal. Let F be the resolution of the identity for N. Then F is a selfadjoint spectral measure, and since $\sigma(N) = \sigma(T) \subset K$, we have

$$\int_K \lambda F(d\lambda) = N = Q\int_K \lambda E(d\lambda)Q^{-1} = \int_K \lambda QE(d\lambda)Q^{-1}.$$

Clearly $QE(\cdot)Q^{-1}$ is a spectral measure on $\mathcal{B}(K)$ and the uniqueness of the integral representation above implies that $QE(\cdot)Q^{-1} = F(\cdot)$, Q.E.D.

We turn now to C^n-operators with real spectrum, for $n \geq 1$. Let $\tau: C^n(R) \to B(X)$ be the C^n-operational calculus for T. For each $x \in X$ and $x^* \in X^*$, the map $f \to x^*\tau(f)x$ is a continuous linear functional on $C^n(R)$ carried by $\sigma(T)$.

Consider any continuous linear functional φ on $C^n(R)$; φ has necessarily compact support, suppφ. If $(\alpha,\beta) \supset \text{supp}\varphi$, then φ induces a continuous linear functional on the Banach space $C^n[\alpha,\beta]$, which we denote also by φ (and whose norm is accordingly denoted by $||\varphi||$, when (α,β) is fixed in our discussion). Consider the Cartesian product $C[\alpha,\beta]^{n+1}$, with elements $\underline{f} = (f_0,f_1,\dots,f_n)$ normed by

$$||\underline{f}|| = \sum_{j=0}^{n} \sup_{[\alpha,\beta]} |f_j|/j!\ .$$

Let A be the linear manifold of all $\underline{f} \in C[\alpha,\beta]^{n+1}$ for which there exists $f \in C^n[\alpha,\beta]$ such that $f_j = f^{(j)}$ $(0 \leq j \leq n)$. Set $\tilde{\varphi}(\underline{f}) = \varphi(f)$, for $\underline{f} \in A$. Since $\text{supp}\varphi \subset (\alpha,\beta)$, $\tilde{\varphi}$ is well-defined, linear, and

$$|\tilde{\varphi}(\underline{f})| \leq ||\varphi||\ ||f||_{C^n[\alpha,\beta]} = ||\varphi||\ ||\underline{f}|| \quad (\underline{f} \in A).$$

By the Hahn-Banach theorem, $\tilde{\varphi}$ extends as a continuous linear functional on $C[\alpha,\beta]^{n+1}$, with norm $\leq ||\varphi||$. Applying the Riesz representation theorem, we obtain $n+1$ regular complex Borel measures $\mu_0,\dots,\mu_n$ on $[\alpha,\beta]$, with total variations $||\mu_j|| \leq ||\varphi||$, such that

$$\tilde{\varphi}(\underline{f}) = \sum_{j=0}^{n} \int_\alpha^\beta f_j d\mu_j \qquad (\underline{f} \in C[\alpha,\beta]^{n+1}).$$

In particular (for $\underline{f} \in A$),

$$\varphi(f) = \sum_{j=0}^{n} \int_{\alpha}^{\beta} f^{(j)} d\mu_j \qquad (f \in C^n [\alpha,\beta]). \tag{13.3.1}$$

In general, φ has many representations of this form, even for (α,β) fixed, because of the non-uniqueness of the Hahn-Banach extension. However uniqueness can be obtained by restricting the kinds of functionals φ and measures μ_j above.

13.4 Definition. The continuous linear functional φ on $C^n(R)$ is said to be singular if it possesses a representation (13.3.1) in which all μ_j $(1 \leq j \leq n)$ are singular with respect to Lebesgue measure. Such a representation is called a singular representation of φ.

Note that there is no restriction on μ_0.

13.5 Lemma. A singular continuous linear functional on $C^n(R)$ has a unique singular representation.

Proof. Using the notation customary in the theory of distributions [46], we must show that if $\sum_{j=0}^{n} \mu_j^{(j)} = 0$ and μ_j $(j \geq 1)$ are singular (with respect to Lebesgue measure), then $\mu_j = 0$ $(0 \leq j \leq n)$ (we wrote μ_j instead of $(-1)^j \mu_j$ to simplify notation).

We have $[\sum_{j=1}^{n} \mu_j^{(j-1)}]' = -\mu_0$, that is, the distribution $\sum_{j=1}^{n} \mu_j^{(j-1)}$ is a primitive of a measure. By Theorem II, p. 54 in [46], this distribution "is" a function of bounded variation g_1 (through the identification customary in the theory of distributions).

Next, if $n \geq 2$, we have

$$[\sum_{j=2}^{n} \mu_j^{(j-2)}]' = -\mu_1+g_1.$$

Again, as a primitive of a measure, $\sum_{j=2}^{n} \mu_j^{(j-2)}$ is a function of bounded variation g_2.

Continuing this process, we obtain finally

$$\mu_n' = -\mu_{n-1}+g_{n-1}, \tag{13.5.1}$$

so that μ_n is a function of bounded variation g_n (that is $\mu_n(dt) = g_n(t)dt$). But μ_n is singular; hence $\mu_n = 0$. Going back to (13.5.1), we get $\mu_{n-1} = g_{n-1}$.

If $n \geq 2$, μ_{n-1} is singular, hence $\mu_{n-1} = 0$. In this manner, we obtain that $\mu_j = 0$ for $j \geq 1$, and hence also $\mu_0 = 0$, since $\mu_0 = -\sum_{j=1}^{n} \mu_j^{(j)}$.

13.6 Definition. Let T be a C^n-operator with real spectrum, and let τ be the $C^n(R)$-operational calculus for T. We say that T is singular if the linear functionals $x^*\tau(\cdot)x$ are singular for all $x \in X$ and $x^* \in X^*$.

13.7 Lemma. Let T be a singular C^n-operator with $\sigma(T) \subset (\alpha,\beta)$, on a reflexive Banach space X. Then there exist a unique ordered set of $n+1$ operator-valued additive set functions $\{F_0,\ldots,F_n\}$ on $\mathcal{B}(R)$, with the following properties:

(1) $x^*F_j(\cdot)x$ is a regular complex Borel measure on R, for each $x \in X$, $x^* \in X^*$, and $j = 0,\ldots,n$ (and therefore F_j are uniformly bounded and strongly σ-additive on $\mathcal{B}(R)$);

(2) F_j $(0 \leq j \leq n)$ are supported by $[\alpha,\beta]$;

(3) for all $x \in X$, $x^* \in X^*$, and $1 \leq j \leq n$, $x^*F_j(\cdot)x$ is singular with respect to Lebesgue measure;

(4) $\tau(f) = \sum_{j=0}^{n} \int_\alpha^\beta f^{(j)} dF_j \qquad (f \in C^n(R))$.

Proof. For each $x \in X$ and $x^* \in X^*$, the singular functional $x^*\tau(\cdot)x$ has the unique singular representation (cf. Lemma 13.5)

$$x^*\tau(f)x = \sum_{j=0}^{n} \int_\alpha^\beta f^{(j)}(t)\mu_j(dt|x,x^*) \quad (f \in C^n(R)) \tag{13.7.1}$$

where μ_j are regular complex Borel measures on $[\alpha,\beta]$, and μ_j are singular (with respect to Lebesgue measure) for $j \geq 1$. Since $x^*\tau(f)x$ is linear in x and x^* for each $f \in C^n(R)$, it follows from the uniqueness of the representation (13.7.1) that $\mu_j(\delta|x,x^*)$ is a bilinear form in x and x^*, for each $\delta \in \mathcal{B}([\alpha,\beta])$ and $0 \leq j \leq n$. In addition (cf. discussion preceding Definition 13.4)

$$|\mu_j(\delta|x,x^*)| \leq ||\mu_j(\cdot|x,x^*)|| \leq ||x^*\tau(\cdot)x|| \leq ||\tau||\ ||x||\ ||x^*||,$$

where $||\tau||$ denotes the norm of the $C^n[\alpha,\beta]$-o.c. for T. Since X is reflexive, there exists a unique operator $E_j(\delta)$ with norm $\leq ||\tau||$, such that

$\mu_j(\delta|x,x^*) = x^*E_j(\delta)x$ for all $x \in X$, $x^* \in X^*$, and $\delta \in \mathcal{B}([\alpha,\beta])$ $(0 \le j \le n)$. Setting $F_j(\delta) = E_j(\delta\cap[\alpha,\beta])$ $(0 \le j \le n)$, the lemma follows.

Notation. If μ is a complex Borel measure on R, we set $\hat{\mu}(t) = \int_R e^{its}d\mu(s)$ (the Fourier-Stieltjes transform of μ).

13.8 Lemma. Let μ_j and ν_j $(0 \le j \le n)$ be regular complex Borel measures with compact support on R, and suppose that $\mu_j-\nu_j$ are singular with respect to Lebesgue measure for $j \ge 1$. Then the relation

$$\sum_{j=0}^{n}(it)^j\hat{\mu}_j(t) = \sum_{j=0}^{n}(it)^j\hat{\nu}_j(t) \qquad (t \in R) \tag{13.8.1}$$

implies that $\mu_j = \nu_j$ for $j = 0,\ldots,n$.

Proof. Let $f_t(s) = e^{its}$ $(s,t \in R)$ and let φ be a continuous linear functional on $C^n(R)$. Set $\hat{\varphi}(t) = \varphi(f_t)$ $(t \in R)$. In terms of the representation (13.3.1), that is,

$$\varphi(f) = \sum_{j=0}^{n}\int_R f^{(j)}(s)d\alpha_j(s) \qquad (f \in C^n(R)), \tag{13.8.2}$$

where α_j are regular complex Borel measures on R supported by $[\alpha,\beta]$, we have

$$\hat{\varphi}(t) = \sum_{j=0}^{n}(it)^j\hat{\alpha}_j(t) \qquad (t \in R). \tag{13.8.3}$$

For $f \in S$, we set here $\hat{f}(t) = \int_R e^{its}f(s)ds$. We then have $\varphi(\hat{f}) = \int_R \hat{\varphi}(t)f(t)dt$. Therefore, if $\hat{\varphi} = 0$, then $\varphi(\hat{f}) = 0$ for all $f \in S$, that is, $\varphi(h) = 0$ for all $h \in S$ (since $\hat{S} = S$). Since φ has compact support, this means that $\varphi = 0$. Thus the map $\varphi \to \hat{\varphi}$ is one-to-one.

Consider now the functional φ defined by (13.8.2) with $\alpha_j = \mu_j-\nu_j$, where μ_j,ν_j satisfy the Lemma's hypothesis (including (13.8.1)). By (13.8.3) and (13.8.1), $\hat{\varphi} = 0$, and therefore $\varphi = 0$ and $\alpha_j = 0$ $(0 \le j \le n)$ by Lemma 13.5, Q.E.D.

13.9 Lemma. Let T be a singular C^n-operator with $\sigma(T) \subset (\alpha,\beta)$ on a reflexive Banach space X, and let F_j $(0 \le j \le n)$ be the operator measures associated with T as in Lemma 13.7. Set $E_j = j!F_j$ and $\hat{E}_j(t) = \int_R e^{its}E_j(ds)$ $(t \in R)$.

Then:

$$\hat{E_k}(t)\hat{E_r}(u) = \hat{E_{k+r}}(t+u) \qquad k+r \le n$$

$$= 0 \qquad k+r > n$$

$(t,u \in R;\ k,r = 0,\dots,n)$.

Proof. Taking $f_t(s) = e^{its}$ in Lemma 13.7(4), we obtain

$$e^{itT} = \sum_{j=0}^{n} \frac{(it)^j}{j!} E_j\hat{}(t) \qquad (t \in R) \quad . \tag{13.9.1}$$

For all $t,u \in R$, we have

$$\sum_{k=0}^{n} \frac{(it)^k}{k!} E_k\hat{}(t)e^{iuT} = e^{i(t+u)T}$$

$$= \sum_{j=0}^{n} \frac{[i(t+u)]^j}{j!} E_j\hat{}(t+u)$$

$$= \sum_{k=0}^{n} \frac{(it)^k}{k!} \sum_{r=0}^{n-k} \frac{(iu)^r}{r!} \hat{E_{k+r}}(t+u) .$$

For $x \in X$, $x^* \in X^*$, and $u \in R$ fixed, the measures $\mu_k(ds) = \frac{1}{k!} x^*E_k(ds)e^{iuT}x$ and

$$\nu_k(ds) = \frac{1}{k!}e^{ius} \sum_{r=0}^{n-k} \frac{(iu)^r}{r!} x^*E_{k+r}(ds)x$$

are singular with respect to Lebesgue measure for $k \ge 1$, by Lemma 13.7(3), and satisfy (13.8.1). Therefore, by Lemma 13.8,

$$e^{ius} \sum_{r=0}^{n-k} \frac{(iu)^r}{r!} E_{k+r}(ds) = E_k(ds)e^{iuT} \qquad (k = 0,\dots,n).$$

Multiply both sides by e^{its} $(t \in R)$, and integrate with respect to s over R; thus, for all $t,u \in R$,

$$\sum_{r=0}^{n-k} \frac{(iu)^r}{r!} \hat{E_{k+r}}(t+u) = E_k\hat{}(t)e^{iuT}$$

$$= \sum_{r=0}^{n} \frac{(iu)^r}{r!} E_k\hat{}(t)E_r\hat{}(u) \tag{13.9.2}$$

for each $k = 0,\dots,n$.

For x, x^*, k, and t fixed, the measures

$$\mu_r(ds) = \frac{1}{r!} e^{its}x^*E_{k+r}(ds)x \qquad 0 \le r \le n-k$$

$$= 0 \qquad n-k < r \le n$$

and $\qquad \nu_r(ds) = \frac{1}{r!} x^* E_k\hat{\ }(t) E_r(ds)x$

$$= \frac{1}{r!}\{[E_k\hat{\ }(t)]^* x^*\} E_r(ds)x$$

are singular with respect to Lebesgue measure for $r \geq 1$ (by Lemma 13.7(3)), and satisfy (13.8.1), by (13.9.2). Applying Lemma 13.8 we obtain $\mu_r = \nu_r$, and therefore $\hat{\mu}_r = \hat{\nu}_r$, for $r = 0,\ldots,n$. This is clearly equivalent to the identities of the lemma.

The following result provides a Jordan decomposition for singular C^n-operators (with real spectrum).

13.10 Theorem. Let T be a bounded linear operator with real spectrum in a reflexive Banach space X. Then T is a singular C^n-operator if and only if it is spectral of type n, and its nilpotent part N and resolution of the identity E are such that the measures $x^*NE(\cdot)x$ $(x \in X, x^* \in X^*)$ are singular with respect to Lebesgue measure.

Proof. The case $n = 0$ is taken care of by Theorem 13.1 (and the result is more generally valid for X weakly complete).

Suppose then that $n \geq 1$.

If T is spectral of type n, then as observed at the beginning of this section, T is of class C^n, and its C^n-o.c. is given by

$$\tau(f) = \sum_{j=0}^{n} \int f^{(j)}(s) N^j E(ds)/j! \qquad (f \in C^n(R)).$$

Thus, for each $x \in X$ and $x^* \in X^*$, the functional $x^*\tau(\cdot)x$ has a representation (13.3.1) with $\mu_j = \mu_j(\cdot|x,x^*) = x^*N^jE(\cdot)x/j!$. For $j \geq 1$, we write

$$\mu_j = [(N^{j-1})^* x^*]NE(\cdot)x/j!.$$

If $x^*NE(\cdot)x$ is singular (with respect to Lebesgue measure) for all x,x^*, then μ_j is singular for $j \geq 1$, and so T is a singular C^n-operator.

Conversely, suppose T is a singular C^n-operator with $\sigma(T) \subset (\alpha,\beta)$, and let E_j be as in Lemma 13.9, $0 \leq j \leq n$.

Taking $k = r = 0$ in Lemma 13.9, we see that

$$E_0\hat{\ }(t+u) = E_0\hat{\ }(t)E_0\hat{\ }(u) \qquad (t,u \in R).$$

It follows that the map $\tau_0: f \to \int f dE_0$ is multiplicative over the algebra of functions f of the form $f(s) = \sum_{j=1}^{m} c_j \exp(it_j s)$ ($c_j \in C$; $s, t_j \in R$; $m = 1,2,\ldots$). Since this algebra is dense in $C[\alpha,\beta]$ and E_0 is uniformly bounded and supported by $[\alpha,\beta]$, we conclude that $\tau_0: C[\alpha,\beta] \to B(X)$ is multiplicative. By (13.9.1) with $t = 0$,

$$\tau_0(1) = I = E_0(R).$$

Thus τ_0 is a $C[\alpha,\beta]$-o.c. for the operator $S = \tau_0(\lambda) = \int \lambda E_0(d\lambda)$, that is, S is of class C, hence spectral of scalar type with resolution of the identity $E = E_0$ (cf. Theorem 13.1).

Next, set $N = E_1(R)$.

For $1 \le j \le n$, take $k = j-1$ and $r = 1$ in Lemma 13.9:

$$E_j\hat{}(t+u) = E_{j-1}\hat{}(t)E_1\hat{}(u) \qquad (1 \le j \le n).$$

For $u = 0$, this gives $E_j\hat{}(t) = E_{j-1}\hat{}(t)N$. Hence

$$E_j\hat{}(t) = E_0\hat{}(t)N^j \qquad (0 \le j \le n).$$

Interchanging the roles of k and r, we obtain also

$$E_j\hat{}(t) = N^j E_0\hat{}(t) \qquad (0 \le j \le n).$$

Equivalently (with $E = E_0$ as before),

$$E_j\hat{}(t) = [E(\cdot)N^j]\hat{}(t) = [N^j E(\cdot)]\hat{}(t) \qquad (t \in R)$$

and therefore, by injectiveness of Fourier-Stieltjes transform

$$E_j = EN^j = N^j E \qquad (0 \le j \le n). \tag{13.10.1}$$

Take now $k = n$, $r = 1$, and $t = u = 0$ in Lemma 13.9. We obtain by (13.10.1)

$$0 = E_n(R)E_1(R) = N^{n+1}.$$

By Lemma 13.7(4) (recall that $E_j = j!F_j$) and (13.10.1),

$$\tau(f) = \sum_{j=0}^{n} \frac{N^j}{j!} \int f^{(j)} dE \qquad (f \in C^n(R)).$$

In particular, with $f(\lambda) = \lambda$, we get

$$T = S + N.$$

Since S is spectral of scalar type, $N^{n+1} = 0$, and N commutes with S

(by (13.10.1)), it follows that T is spectral of type n. In addition, for all $x \in X$ and $x^* \in X^*$, the measure $x^*NE(\cdot)x = x^*E_1(\cdot)x$ (cf. (13.10.1)) is singular with respect to Lebesgue measure, by Lemma 13.7(3).

13.11 Corollary. Let T be a bounded linear operator on a reflexive Banach space, whose spectrum is a finite subset of R. Then T is of class C^n if and only if it is spectral of type n.

Proof. (for the "only if" part) Let T be of class C^n, and let τ be its $C^n(R)$-o.c. For each $x \in X$ and $x^* \in X^*$, $x^*\tau(\cdot)x$ is a distribution of order $\leq n$ with support $\sigma(T) = \{p_1,\dots,p_m\} \subset R$ (cf. Proposition 4.4). By [46; pp. 99-100], there exist $M > 0$ and $c_{ij} = c_{ij}(x,x^*) \in C$ $(i = 1,\dots,m; j = 0,\dots,n)$ such that

$$x^*\tau(f)x = \sum_{i,j} c_{ij}f^{(j)}(p_i) \qquad (f \in C^n(R))$$

and $|c_{ij}(x,x^*)| \leq M||x|| \, ||x^*|| \qquad (x \in X, x^* \in X^*)$.

Thus $x^*\tau(\cdot)x$ has a representation (13.3.1) with $\mu_j = \sum_{i=1}^{m} c_{ij}\delta_{p_i}$ $(0 \leq j \leq n)$, where δ_{p_i} denotes the delta measure at p_i. Since all μ_j are singular with respect to Lebesgue measure, T is a singular C^n-operator, and the conclusion follows from Theorem 13.10.

13.12 Example. When $\sigma(T)$ is an infinite (real) sequence, the conclusion of Corollary 13.11 is no longer true in general, as we shall see in the following example.

Let $X = \sum_{n=1}^{\infty} \oplus H_n$, where H_n is a copy of the 2-dimensional Hilbert space C^2. Consider on $_n$ the projections $E_{\pm n}$ defined by the matrices $\begin{pmatrix} 1 & n \\ 0 & 0 \end{pmatrix}$ and $\begin{pmatrix} 0 & -n \\ 0 & 1 \end{pmatrix}$, $n = 1,2,\dots$. Clearly,

$$E_{-n}E_n = E_nE_{-n} = 0; \; E_{\pm n}^2 = E_{\pm n}; \text{ and}$$

$$E_n + E_{-n} = I.$$

Let $T_n = (1/n)(E_n - E_{-n})$ and

$$T = \sum_{n=1}^{\infty} \oplus T_n.$$

Then $\sigma(T) = \{0, \pm 1/n;\ n = 1,2,\dots\}$ and T is of class C^1: its C^1-operational calculus is given by $\tau(f) = \sum_{n=1}^{\infty} \oplus \tau(f)_n$, where

$$\tau(f)_n = f(1/n)E_n + f(-1/n)E_{-n} \qquad (f \in C^1).$$

The continuity of τ follows from the mean value theorem, which gives the estimate

$$||\tau(f)|| \leq \max_{\Delta}|f| + 2\max_{\Delta}|f'| \qquad (f \in C^1),$$

where $\Delta = [-1,1]$.

The Riesz-Dunford projection associated with the isolated point $1/n$ of the spectrum is

$$E(\{1/n\}) = 0 \oplus \ldots \oplus E_n \oplus 0 \oplus \ldots \qquad (n = \pm 1, \pm 2, \ldots),$$

and has norm $\geq |n|$. Since the projections $E(\{1/n\})$ are not uniformly bounded, T is not spectral.

Yet, when $\sigma(T) \subset R$ has Lebesgue measure 0, a "local version" of the Jordan decomposition theorem can be obtained. We start with some notation.

13.13 Definition. Let m be a non-negative integer, and let $T \in B(X)$ have real spectrum. We set

$$||f||_{m,T} = \sum_{j=0}^{m} \max_{\sigma(T)} |f^{(j)}|/j! \qquad (f \in C^m(\sigma(T))).$$

As before, P denotes the algebra of all polynomials over C (restricted to R). Let

$$P_{m,T} = \{p \in P;\ ||p||_{m,T} \leq 1\}$$

$$|x|_{m,T} = \sup\{||p(T)x||;\ p \in P_{m,T}\} \qquad (x \in X)$$

$$D_{m,T} = \{x \in X;\ |x|_{m,T} < \infty\}.$$

As T will be fixed throughout the following discussion, we shall omit the index T in the preceding notation (using $||f||_m$, $|x|_m$, and D_m instead).

13.14 Lemma. (1) $(D_m, |\cdot|_m)$ is a Banach space.

(2) $D_0 \subset D_1 \subset D_2 \subset \ldots$

(3) D_m is invariant for any $V \in B(X)$ which commutes with T, and

$$|Vx|_m \leq ||V||\ |x|_m \qquad (x \in D_m).$$

(4) $|p(T)x|_m \leq ||p||_m |x|_m \qquad (p \in P;\ x \in D_m)$.

Proof. Since $|\cdot|_m$ is clearly homogeneous and subadditive, and majorizes $||\cdot||$ (since $1 \in P_m$), $(D_m, |\cdot|_m)$ is a normed space. Suppose $\{x_k\} \subset D_m$ is $|\cdot|_m$-Cauchy. It is then $||\cdot||$-Cauchy; let $x = \lim x_k$ in X. Given $\varepsilon > 0$, let k_0 be such that

$$|x_k - x_h|_m < \varepsilon/2 \qquad (k,h > k_0).$$

Then $||p(T)x_k - p(T)x_h|| < \varepsilon/2 \quad (p \in P_m)$.

Let $h \to \infty$; then

$$||p(T)(x_k - x)|| \leq \varepsilon/2 \qquad (k > k_0;\ p \in P_m),$$

that is $|x_k - x|_m < \varepsilon \ (k > k_0)$.

Hence $x \in D_m$ and $x_k \to x$ in $(D_m, |\cdot|_m)$. This proves (1).

For x fixed, $|x|_m$ is a non-increasing function of m $(m = 0,1,2,\ldots)$, and (2) follows.

If $V \in B(X)$ commutes with T, it commutes also with $p(T)$ for all $p \in P$. Therefore, for all $p \in P_m$,

$$||p(T)Vx|| = ||Vp(T)x|| \leq ||V||\ ||p(T)x|| \leq ||V||\ |x|_m,$$

and (3) is proved.

Fix $x \in D_m$ and $p \in P$. For all $q \in P_m$,

$$\begin{aligned} ||q(T)[p(T)x]|| &= ||(qp)(T)x|| \leq ||qp||_m |x|_m \\ &\leq ||q||_m ||p||_m |x|_m \leq ||p||_m |x|_m, \end{aligned}$$

and (4) follows.

In order to avoid repetition, we set the following

Standing hypothesis. X is a reflexive Banach space; $T \in B(X)$ has real spectrum, and the (linear) Lebesgue measure of $\sigma(T)$ is zero.

13.15 Lemma. There exists a uniquely determined family $\{E_{j,m}(\delta);\ m = 0,1,2,\ldots;\ j = 0,\ldots,m;\ \delta \in B(R)\}$ of linear transformations of X with the following properties:

(a) $E_{j,m}(\delta)$ has domain D_m;

(b) for each $x \in D_m$, $E_{j,m}(\cdot)x$ is a regular, uniformly bounded, strongly countably additive vector measure supported by $\sigma(T)$; and

(c) $p(T)x = \sum_{j=0}^{m} (1/j!)\int_{\sigma(T)} p^{(j)}(t)E_{j,m}(dt)x$

for all $x \in D_m$ and $p \in P$.

Proof. Let $C(\sigma(T))^{m+1}$ be the Cartesian product of $m+1$ copies of $C(\sigma(T))$, with elements $\underline{f} = (f_0, f_1, \ldots, f_m)$ normed by

$$||\underline{f}|| = \sum_{j=0}^{m} \sup_{\sigma(T)} |f_j|/j! \ .$$

Let A_m be the linear manifold of $C(\sigma(T))^{m+1}$ consisting of all $\underline{f}$ with $f_j = p^{(j)}|\sigma(T)$ for some $p \in P$ $(j = 0,\ldots,m)$.

Fix m, $x \in D_m$, and $x^* \in X^*$, and consider the map

$$\pi_m(x,x^*): \underline{f} \in A_m \to x^*p(T)x$$

where $f_j = p^{(j)}|\sigma(T)$.

By Lemma 13.14(4),

$$|\pi_m(x,x^*)\underline{f}| \le ||x^*||\ ||p(T)x|| \le ||x^*||\ |p(T)x|_m$$

$$\le ||x^*||\ |x|_m ||p||_m = ||x^*||\ |x|_m ||\underline{f}||.$$

Thus, $\pi_m(x,x^*)$ is well defined, linear and bounded (by $||x^*||\ |x|_m$) on A_m.

By the Hahn-Banach and the Riesz representation theorems, there exist $m+1$ regular complex Borel measures $\mu_{j,m}(\cdot|x,x^*)$ $(j = 0,\ldots,m)$ on $B(R)$, which are supported by $\sigma(T)$ and satisfy

$$\pi_m(x,x^*)\underline{f} = \sum_{j=0}^{m} (1/j!)\int_{\sigma(T)} p^{(j)}(t)\mu_{j,m}(dt|x,x^*) \qquad (\underline{f} \in A_m)$$

and

$$\text{var}\ \mu_{j,m}(\cdot|x,x^*) \le ||x^*||\ |x|_m \qquad (j = 0,\ldots,m). \tag{13.15.1}$$

The continuous linear functional

$$\tilde{\pi}_m(x,x^*): f \to \sum_{j=0}^{m} (1/j!)\int_R f^{(j)}(t)\mu_{j,m}(dt|x,x^*)$$

on $C^m(R)$ is singular (cf Definition 13.4) since $\mu_{j,m}$ are supported by the Lebesgue null set $\sigma(T)$. By Lemma 13.5, the measures $\mu_{j,m}$ are uniquely determined by the functional $\tilde{\pi}_m$, which is itself uniquely determined by its action on P, since P is dense in $C^m[\alpha,\beta]$, where (α,β) is an interval containing the support $\sigma(T)$ of $\tilde{\pi}_m$. Since

$$\tilde{\pi}_m(x,x^*)p = \pi_m(x,x^*)\underline{f} \quad (\text{where } f_j = p^{(j)}|\sigma(T))$$
$$= x^*p(T)x$$

we therefore have the unique representation

$$x^*p(T)x = \sum_{j=0}^{m}(1/j!)\int_{\sigma(T)} p^{(j)}(t)\mu_{j,m}(dt|x,x^*) \qquad (p \in P). \qquad (*)$$

The linearity of the left hand side in x^* implies consequently the linearity of $\mu_{j,m}(\delta|x,x^*)$ in x^* (with all other parameters fixed, $\delta \in B(R)$). Since X is reflexive, there exists a unique element of X, which we denote by $E_{j,m}(\delta)x$, with norm $\leq |x|_m$, such that $\mu_{j,m}(\delta|x,x^*) = x^*E_{j,m}(\delta)x$ for a.p.i. (all parameters involved). Necessarily, $E_{j,m}(\delta)$ is a linear transformation (with domain D_m). Now (b) follows from the Orlicz-Banach-Pettis theorem, and (c) is a rewriting of (*).

13.16 Lemma. If $V \in B(X)$ commutes with T, then V commutes with $E_{j,m}(\delta)$ for a.p.i. (that is $VE_{j,m}(\delta)x = E_{j,m}(\delta)Vx$ for all $x \in D_m$).

Proof. If V commutes with T, then $VD_m \subset D_m$ by Lemma 13.14(3) $(m = 0,1,\ldots)$. Therefore, for all $x \in D_m$ and $p \in P$, we have by Lemma 13.15(c)

$$\sum_{j=0}^{m}(1/j!)\int_{\sigma(T)} p^{(j)}(t)VE_{j,m}(dt)x = Vp(T)x$$
$$= p(T)Vx = \sum_{j=0}^{m}(1/j!)\int_{\sigma(T)} p^{(j)}(t)E_{j,m}(dt)Vx.$$

By density of P in $C^m[\alpha,\beta]$ (where (α,β) is any open interval containing $\sigma(T)$) and uniqueness of singular representations (Lemma 13.5), it follows that $VE_{j,m}(\delta)x = E_{j,m}(\delta)Vx$ for all $x \in D_m$, $j = 0,\ldots,m$, and $\delta \in B(R)$.

Let $\underline{B}(\sigma(T))$ denote the space of all complex Borel functions on R which are bounded on $\sigma(T)$. By Lemma 13.15(b), the integrals $\int_{\sigma(T)} f(t)E_{j,m}(dt)x$ are well defined elements of X, for all $f \in \underline{B}(\sigma(T))$, $x \in D_m$, and $0 \leq j \leq m < \infty$. Actually, these are elements of D_m, as it follows in particular from the next lemma.

13.17 Lemma. For all $f \in \underline{B}(\sigma(T))$, $x \in D_m$, and $0 \le j \le m < \infty$, we have

$$\left|\int_{\sigma(T)} f(t)E_{j,m}(dt)x\right|_m \le \sup_{\sigma(T)} |f|\ |x|_m.$$

Proof. Fix f,x,j, and m as above, and set $y = \int_{\sigma(T)} f(t)E_{j,m}(dt)x$. With notation as in the proof of Lemma 13.15, it follows from Lemma 13.16, (13.15.1), and Lemma 13.14(4) that for all $p \in P_m$ and $x \in X^*$,

$$\begin{aligned}|x^*p(T)y| &= \left|\int_{\sigma(T)} f(t)x^*E_{j,m}(dt)p(T)x\right| \\ &= \left|\int_{\sigma(T)} f(t)\mu_{j,m}(dt|p(T)x,x^*)\right| \\ &\le \sup_{\sigma(T)} |f| \operatorname{var} \mu_{j,m}(\cdot|p(T)x,x^*) \\ &\le \sup_{\sigma(T)} |f|\ ||x^*||\,|p(T)x|_m \\ &\le \sup_{\sigma(T)} |f|\ ||x^*||\ |x|_m, \qquad \text{Q.E.D.}\end{aligned}$$

In particular, $E_{j,m}(\delta)D_m \subset D_m$, and in fact

$$|E_{j,m}(\delta)x|_m \le |x|_m \tag{13.17.1}$$

for all $\delta \in B(R)$, $x \in D_m$, and $0 \le j \le m < \infty$.

Taking $p = 1$ in Lemma 13.15(c), we obtain

$$E_{o,m}(R) = I/D_m . \tag{13.17.2}$$

Set

$$\hat{E}_{j,m}(s)x = \int_{\sigma(T)} e^{its}E_{j,m}(dt)x \qquad (s \in R;\ x \in D_m;\ 0 \le j \le m < \infty).$$

By Lemma 13.17, $\hat{E}_{j,m}(s)\colon D_m \to D_m$ and

$$|\hat{E}_{j,m}(s)x|_m \le |x|_m \qquad (x \in D_m).$$

13.18 Lemma. For all $0 \le j,h \le m < \infty$ and $s,u \in R$,

$$\begin{aligned}\hat{E}_{j,m}(s)\hat{E}_{h,m}(u) &= \hat{E}_{j+h,m}(s+u) && \text{for } j+h \le m \\ &= 0 && \text{for } j+h > m.\end{aligned}$$

Proof. Since $\sigma(T)$ has a connected complement in the Riemann sphere, it follows from Lemma 13.15(c) and polynomial approximation that

$$f(T)x = \sum_{j=0}^{m} (1/j!)\int_{\sigma(T)} f^{(j)}(t)E_{j,m}(dt)x \tag{13.18.1}$$

for all $f \in H(\sigma(T))$, $x \in D_m$, $m = 0,1,2,\ldots$.

In particular,

$$e^{isT}x = \sum_{j=0}^{m} \frac{(is)^j}{j!} \hat{E}_{j,m}(s)x \qquad (s \in R\ ,\ x \in D_m). \qquad (13.18.2)$$

Using the group property of e^{isT} and the invariance of D_m under the group, we obtain after some rearrangement

$$\sum_{h=0}^{m} (is)^h \int_{\sigma(T)} e^{ist}[e^{iut} \frac{1}{h!} \sum_{j=0}^{m-h} \frac{(iu)^j}{j!} E_{j+h,m}(dt)x]$$

$$= \sum_{h=0}^{m} (is)^h \int_{\sigma(T)} e^{ist}[\frac{1}{h!} E_{h,m}(dt)e^{iuT}x]$$

for all $s,u \in R$ and $x \in D_m$.

Since $\sigma(T)$ is a Lebesgue null set, it follows from Lemma 13.8 (applied with u fixed and s variable) that for all $s,u \in R$, $x \in D_m$, and $0 \le h \le m$,

$$\int_{\sigma(T)} e^{ist}e^{iut} \sum_{j=0}^{m-h} \frac{(iu)^j}{j!} E_{j+h,m}(dt)x$$

$$= \int_{\sigma(T)} e^{ist}E_{h,m}(dt)e^{iuT}x$$

$$= \hat{E}_{h,m}(s)e^{iuT}x = e^{iuT}\hat{E}_{h,m}(s)x$$

$$= \sum_{j=0}^{m} \frac{(iu)^j}{j!} \int_{\sigma(T)} e^{iut}E_{j,m}(dt)\hat{E}_{h,m}(s)x,$$

where we used Lemma 13.16, the invariance of D_m under $\hat{E}_{h,m}(s)$, and (13.18.2).

Rewrite the extreme left in the form

$$\sum_{j=0}^{m-h} \frac{(iu)^j}{j!} \int_{\sigma(T)} e^{iut}[e^{ist}E_{j+h,m}(dt)x]$$

and apply Lemma 13.8 with s fixed and u variable. We obtain

$$E_{j,m}(dt)\hat{E}_{h,m}(s)x = e^{ist}E_{j+h,m}(dt)x \qquad j+h \le m$$
$$= 0 \qquad j+h > m$$

for all $s,t \in R$ and $x \in D_m$.

Multiplying by e^{iut} and integrating with respect to t over $\sigma(T)$, we get the identities of the lemma.

13.19 <u>Definition</u>. Let $W \subset X$ be a linear manifold. We denote by $T^{\cdot}(W)$ the algebra of all linear transformations of X with domain W and range contained in W.

A <u>generalized spectral measure on W</u> is a map $\delta \to E(\delta)$ of $\mathcal{B}(R)$ into $T(W)$ with the following properties:

(i) $E(R) = I|W$;

(ii) for each $x \in W$, $E(\cdot)x$ is a bounded, regular, strongly countably additive vector measure on $\mathcal{B}(R)$, carried by some compact set $K \subset R$,

(iii) if we set

$$S(f)x = \int_K f(t)E(dt)x \quad (x \in W,\ f \in C(K)),$$

then $S(\cdot)$ is an algebra homomorphism of $C(K)$ into $T(W)$.

13.20 <u>Lemma</u>. Fix $m = 0,1,2,\ldots$, and set $E_m = E_{o,m}$. Then E_m is a generalized spectral measure on D_m.

<u>Proof</u>. Properties (i) and (ii) (with $W = D_m$ and $K = \sigma(T)$) and the fact that $S(\cdot) = S_m(\cdot)$ is a linear map of $C(K)$ into $T(D_m)$ are already known. By Lemma 13.18 with $j = h = 0$, we have

$$E_m\hat{}(s)E_m\hat{}(u) = E_m\hat{}(s+u) \qquad (s,u \in R).$$

It follows that the map $f \to S_m(f)$ is multiplicative when restricted to the set A of functions f of the form $f(t) = \sum_{j=1}^{k} c_j \exp(is_j t)$ $(c_j \in C;\ s_j \in R;\ k = 1,2,\ldots)$. By the Stone-Weierstrass theorem, A is dense in $C(\sigma(T))$. Let $f,g \in C(\sigma(T))$, and let $f_n, g_n \in A$ converge to f,g respectively, uniformly on $\sigma(T)$. For $x \in D_m$, we have

$$||[S(f)S(g) - S(fg)]x|| \leq ||S(f-f_n)S(g)x|| + ||S(f_n)S(g-g_n)x||$$
$$+ ||S(f_ng_n-fg)x|| = A + B + C.$$

By Lemma 13.17, as $n \to \infty$,

$$A \leq \sup_{\sigma(T)} |f-f_n|\ |S(g)x|_m \to 0;$$

$$B \leq \sup_{\sigma(T)} |f_n|\ |S(g-g_n)x|_m \leq \sup_{\sigma(T)} |f_n| \sup_{\sigma(T)} |g-g_n|\ |x|_m \to 0;$$

and

$$C \leq \sup_{\sigma(T)} |f_ng_n-fg|\ |x|_m \to 0.$$

Hence $S(\cdot)$ is multiplicative on $C(\sigma(T))$, Q.E.D.

We set

$$S_m x = S_m(t)x = \int_{\sigma(T)} tE_m(dt)x \qquad (x \in D_m)$$

and

$$N_m = E_{1,m}(\mathbb{R}). \qquad (13.20.1)$$

By Lemma 13.15(c) with $p(t) = t$, we have

$$T|D_m = S_m + N_m. \qquad (13.20.2)$$

13.21 Lemma. (i) N_m commutes with $S_m(f)$ for all $f \in C(\sigma(T))$;

(ii) $E_{j,m}(\delta) = E_m(\delta)N_m^{\,j} \qquad (\delta \in \mathcal{B}(\mathbb{R});\ j = 1,\dots,m)$;

(iii) $N_m^{\,m+1} = 0$.

Proof. Taking $h = 1$ and $u = 0$ in Lemma 3.18, we get

$$\begin{aligned} \hat{E}_{j,m}(s)N_m &= \hat{E}_{j+1,m}(s) && j \leq m-1 \\ &= 0 && j = m. \end{aligned}$$

Hence

$$\hat{E}_{j,m}(s) = \hat{E}_m(s)N_m^{\,j} \qquad (1 \leq j \leq m) \qquad (13.21.1)$$

and

$$\hat{E}_m(s)N_m^{\,m+1} = \hat{E}_{m,m}(s)N_m = 0$$

for all $s \in \mathbb{R}$. Taking $s = 0$ in the last relations, we obtain (iii) (since $\hat{E}_m(0) = E_m(\mathbb{R}) = I|D_m$). Interchanging the roles of j and h in the above argument, we obtain similarly that

$$\hat{E}_{j,m}(s) = N_m^{\,j}\hat{E}_m(s) \qquad (1 \leq j \leq m;\ s \in \mathbb{R}).$$

Together with (13.21.1), this shows that N_m commutes with $S_m(f)$ for all $f \in A$. Using Lemma 13.17 and the density of A in $C(\sigma(T))$, we conclude that N_m commutes with $S_m(f)$ for all $f \in C(\sigma(T))$. Finally, (ii) follows from (13.21.1) and the uniqueness theorem for Fourier-Stieltjes transforms of regular Borel measures, Q.E.D.

Using Lemma 13.21(ii), we may rewrite relation (c) of Lemma 13.15 in the form

$$p(T)x = \sum_{j=0}^{m} (1/j!)\int_{\sigma(T)} p^{(j)}(t)E_m(dt)N_m^{\,j}x \qquad (13.21.2)$$

$(p \in P;\ x \in D_m;\ m = 0,1,2,\dots)$.

The right hand side makes sense for any p in $C^m(\sigma(T))$ (functions of class C^m in a neighborhood of $\sigma(T)$), and not merely for polynomials p.

We let

$$\tau_m(f)x = \sum_{j=0}^{m}(1/j!)\int_{\sigma(T)} f^{(j)}(t)E_m(dt)N_m{}^j x$$

$$= \sum_{j=0}^{m}(1/j!)S_m(f^{(j)})N_m{}^j x \qquad (x \in D_m;\ f \in C^m(\sigma(T))).$$

By Lemma 13.17, $\tau_m: C^m(\sigma(T)) \to T(D_m)$ satisfies the "continuity" requirement

$$|\tau_m(f)x|_m \leq ||f||_m|x|_m \qquad (x \in D_m;\ f \in C^m(\sigma(T))) \qquad (13.21.3)$$

(cf. Definition 13.13).

The map τ_m is trivially linear; its multiplicativity follows from that of $S_m(\cdot)$ (Lemma 13.20) and Lemma 13.21: for $f,g \in C^m(\sigma(T))$,

$$\tau_m(fg) = \sum_{j=0}^{m}\sum_{h=0}^{j}\frac{1}{h!(j-h)!}S_m(f^{(h)})S_m(g^{(j-h)})N_m{}^j$$

$$= \sum_{h=0}^{m}(1/h!)S_m(f^{(h)})\sum_{k=0}^{m-h}(1/k!)S_m(g^{(k)})N_m{}^{k+h}.$$

Since $N_m{}^{m+1} = 0$, the inner summation may be extended up to m, so that

$$\tau_m(fg) = \sum_{h=0}^{m}(1/h!)S_m(f^{(h)})\tau_m(g)N_m{}^h.$$

By Lemma 13.21(i), N_m commutes with $\tau_m(g)$. Hence $\tau_m(fg) = \tau_m(f)\tau_m(g)$.

By (13.21.3), τ_m is a $C^m(\sigma(T))$-operational calculus for T <u>in the Banach space</u> $(D_m,|\cdot|_m)$, that is $T|D_m$ is of class $C^m(\sigma(T))$ in <u>that</u> Banach space.

We collect the information obtained in the preceding lemmas (and some more) in the following

13.22 <u>Theorem.</u> Let T be a bounded linear operator on the reflexive Banach space X. Suppose $\sigma(T) \subset R$ has (linear) Lebesgue measure zero. For $m = 0,1,2,\ldots$, there exist transformations $S_m, N_m \in T(D_m)$ such that

(1) $T|D_m = S_m + N_m$;

(2) $S_mN_m = N_mS_m$;

(3) $N_m{}^{m+1} = 0$; and

(4) $p(S_m)x = \int_{\sigma(T)} p(t)E_m(dt)x \qquad (x \in D_m,\ p \in P)$,

where E_m is a generalized spectral measure on D_m supported by $\sigma(T)$ and commuting with every $V \in B(X)$ which commutes with T.

$T|D_m$ is an operator of class $C^m(\sigma(T))$ on the Banach space $(D_m, |\cdot|_m)$, and its $C^m(\sigma(T))$-o.c. on that space is given by

$$\tau_m(f)x = \sum_{j=0}^{m} (1/j!)\int_{\sigma(T)} f^{(j)}(t)E_m(dt)N_m{}^j x$$

$(x \in D_m;\ f \in C^m(\sigma(T)))$.

The m-th Jordan decomposition (1)-(4) is "maximal-unique", meaning that if W is an invariant linear manifold for T for which (1)-(4) are valid with W replacing D_m, then $W \subset D_m$, and the transformations S,N and $E(\delta)$ $(\delta \in \mathcal{B}(R))$ corresponding to W are the restriction to W of S_m, N_m and $E_m(\delta)$ respectively.

Proof. It remains to prove the "maximality-uniqueness" assertion. Let W,S,N, and $E(\cdot)$ be as described in the statement of the theorem. By Properties (1) and (2) of S and N,

$$T^n x = \sum_{j=0}^{n} \binom{n}{j} S^{n-j}N^j x$$

for all $x \in W$ and $n = 0,1,\ldots$ $(S^0 = N^0 = I$ by definition).

Let $p(t) = \sum_{n=0}^{q} c_n t^n$ $(c_n \in C)$. Then, for $x \in W$,

$$p(T)x = \sum_{j=0}^{q} \sum_{n=j}^{q} c_n \binom{n}{j} S^{n-j}N^j x$$

$$= \sum_{j=0}^{m} (1/j!)\int_{\sigma(T)} \sum_{n=j}^{q} c_n n(n-1)\ldots(n-j+1)t^{n-j}E(dt)N^j x$$

$$= \sum_{j=0}^{m} (1/j!)\int_{\sigma(T)} p^{(j)}(t)E(dt)N^j x \qquad (13.22.1)$$

by properties (3) and (4).

Since the vector measure $E(\cdot)w$ is bounded for each fixed $w \in W$, and since $NW \subset W$, we have

$$M(x) = \sup_{0\le j\le m}\ \sup_{\delta\in\mathcal{B}(R)} ||E(\delta)N^j x|| < \infty$$

for each $x \in W$.

Therefore (cf. [5], Lemma III.1.5)

$$||p(T)x|| \le 4M(x)||p||_m \qquad (x \in W;\ p \in P)$$

and $\qquad |x|_m \le 4M(x) \qquad (x \in W)$.

This proves that $W \subset D_m$.

By (13.21.2) and (13.22.1), and the density of P in $C^m[\alpha,\beta]$ (where $(\alpha,\beta) \supset \sigma(T)$), we have (for $x \in W$ fixed)

$$\sum_{j=0}^{m} (1/j!)\int_{\sigma(T)} f^{(j)}(t)E_m(dt)N_m^{\,j}x = \sum_{j=0}^{m} (1/j!)\int_{\sigma(T)} f^{(j)}(t)E(dt)N^jx$$

for all $f \in C^m[\alpha,\beta]$. Since $\sigma(T)$ is a Lebesgue null set, it follows from the uniqueness of singular representations (lemma 13.5) that

$$E(\delta)N^jx = E_m(\delta)N_m^{\,j}x \qquad (\delta \in \mathcal{B}(R);\ 0 \le j \le m).$$

In particular $(j = 0)$, $E(\delta)x = E_m(\delta)x$ for all $\delta \in \mathcal{B}(R)$ and $x \in W$; hence $Sx = S_mx$ by Property (4) and $Nx = N_mx$ by Property (1), for all $x \in W \subset D_m$. This completes the proof of Theorem 13.22.

The existence and uniqueness theorem 13.22 justifies the following terminology. For $m = 0,1,\ldots$, $D_m = D_{m,T}$ is the <u>m-th Jordan manifold</u> of T; S_m, N_m, and E_m are respectively the <u>scalar part</u> (or semi-simple part), the <u>nilpotent part</u>, and the <u>resolution of the identity</u> of T on D_m. Since $T|D_0 = S_0$ (cf. Theorem 13.22(1),(3)), we refer to D_0 as the <u>semi-simplicity manifold</u> of T.

Keeping in mind the usual definition of a resolution of the identity, the following proposition is of some interest. The operator T is as in the preceding theorem (that is, $\sigma(T) \subset R$ is a Lebesgue null set!), and we fix $m \ge 0$.

13.23 <u>Proposition</u>. If the nilpotent part (or the scalar part) of T on D_m is closable, then $E_m(\delta)$ commutes with S_m, N_m and $S_m(f)$ for all $\delta \in \mathcal{B}(R)$ and $f \in C(\sigma(T))$, and moreover

$$E_m(\delta \cap \varepsilon) = E_m(\delta)E_m(\varepsilon) \qquad (\delta,\varepsilon \in \mathcal{B}(R)).$$

<u>Proof</u>. Fix $\delta \in \mathcal{B}(R)$, $x \in D_m$, and $f \in C(\sigma(T))$. Recalling that $S_m(f)x \in D_m$ (cf. Lemma 13.20), we may choose a finite positive measure λ such that the vector measures $E_m(\cdot)x$ and $E_m(\cdot)S_m(f)x$ are both λ-continuous (cf. [5; p. 321]). Let χ_δ denote the characteristic function of δ, and choose $g_n \in L(R)$ such that $|g_n| \le 1$ and $g_n \to \chi_\delta$ almost everywhere $[\lambda]$ on $\sigma(T)$. Then $|g_nf| \le |f|$ and $g_nf \to \chi_\delta f$ a.e $[\lambda]$ on $\sigma(T)$. Applying the dominated convergence theorem for vector

measures (cf. [5; p. 328]), we obtain as $n \to \infty$:

$$S_m(g_n)S_m(f)x \to E_m(\delta)S_m(f)x \tag{13.23.1}$$

and

$$S_m(g_n f)x \to S_m(\chi_\delta f)x. \tag{13.23.2}$$

Since $S_m(\cdot)$ is multiplicative (cf. Lemma 13.20), we conclude that

$$S_m(\chi_\delta f) = E_m(\delta)S_m(f) \qquad (f \in C(\sigma(T));\ \delta \in \mathcal{B}(R)). \tag{13.23.3}$$

Let $y_n = S_m(g_n)x - E_m(\delta)x$. Then $y_n \in D_m$ and $y_n \to 0$ (take $f \equiv 1$ in (13.23.1)). Also, as $n \to \infty$,

$$S_m y_n = S_m(g_n(t)t)x - S_m E_m(\delta)x \to E_m(\delta)S_m x - S_m E_m(\delta)x$$

by (13.23.2) and (13.23.3) with $f(t) \equiv t$.

Since S_m is closable (either by hypothesis, or because $S_m = T|D_m - N_m$ where $T|D_m$ is bounded and N_m is closable), it follows that $E_m(\delta)S_m x = S_m E_m(\delta)x$. Thus $E_m(\delta)$ commutes with S_m, and since it commutes with T (cf. Lemma 13.16), it commutes also with $N_m = T|D_m - S_m$. Now $E_m(\delta)$ commutes with $p(S_m)$ for all $p \in P$; by Lemma 13.17 and the density of P in $C(\sigma(T))$, it follows that $E_m(\delta)$ commutes with $S_m(f)$ for all $f \in C(\sigma(T))$. Thus, by (13.23.3), $S_m(\chi_\delta f) = S_m(f)E_m(\delta)$, that is

$$\int_{\sigma(T)} f(t)\chi_\delta(t)E_m(dt)x = \int_{\sigma(T)} f(t)E_m(dt)E_m(\delta)x$$

for all $f \in C(\sigma(T))$ and $x \in D_m$.

By the uniqueness of the Riesz representation, we get

$$\chi_\delta(t)E_m(dt)x = E_m(dt)E_m(\delta)x,$$

hence $E_m(\varepsilon \cap \delta)x = E_m(\varepsilon)E_m(\delta)x$ for all $\varepsilon, \delta \in \mathcal{B}(R)$ and $x \in D_m$, Q.E.D.

Since $N_0 = 0$ is trivially closable, the conlcusions of Proposition 13.23 are valid on the semi-simplicity manifold D_0, that is $E_0(\delta)$ commutes wtih $S_0(f)$ for all $\delta \in \mathcal{B}(R)$ and $f \in C(\sigma(T))$, and $E_0(\delta \cap \varepsilon) = E_0(\delta)E_0(\varepsilon)$ $(\delta, \varepsilon \in \mathcal{B}(R))$. For $m = 0$, the hypothesis that $\sigma(T)$ be a Lebesgue null set is superfluous. Indeed, it was used only to insure the uniqueness of the integral representations met along the way; however, for $m = 0$, the uniqueness is provided by the Riesz representation theorem, without any assumption on $\sigma(T)$. Actually, $\sigma(T)$ may be replaced throughout the theory by any compact subset K of R containing $\sigma(T)$. Due to the particular importance of the special case $m = 0$, we restate for it our definitions and results.

13.24 Definition. Let $T \in B(X)$ have real spectrum. For $x \in X$, set

$$|x|_0 = |x|_{0,T} = \sup\{\|p(T)x\|;\ p \in P,\ \sup_{\sigma(T)}|p| \leq 1\}.$$

The semi-simplicity manifold for T is the set

$$D_0 = \{x \in X;\ |x|_0 < \infty\}.$$

If W is a linear manifold in X, a spectral measure on W is a map $\delta \to E(\delta)$ of $\mathcal{B}(R)$ into $T(W)$ such that:

(i) $E(R) = I|W$;

(ii) for each $x \in W$, $E(\cdot)x$ is a regular, strongly countably additive vector measure on $\mathcal{B}(R)$; and

(iii) $E(\delta \cap \varepsilon) = E(\delta)E(\varepsilon)$ for all $\delta, \varepsilon \in \mathcal{B}(R)$.

By [5; Corollary III.4.5], $E(\cdot)x$ is necessarily bounded (with a bound depending of course on $x \in W$).

13.25 Theorem. Let X be reflexive Banach space and let $T \in B(X)$ have real spectrum. Then there exists a spectral measure E on the semi-simplicity manifold D_0 for T, supported by $\sigma(T)$ and commuting with each $V \in B(X)$ which commutes with T, such that

(1) $$p(T)x = \int_{\sigma(T)} p(t)E(dt)x \qquad (x \in D_0,\ p \in P).$$

E is "maximal-unique", that is, if E' is a spectral measure on W (even with (iii) omitted in Definition 13.24) satisfying (1) for all $x \in W$, then $E'(\delta) \subset E(\delta)$ for all $\delta \in \mathcal{B}(R)$.

14. Local analysis.

Theorems 13.22 and 13.25 are examples of what we shall mean here by "local analysis".

14.1 Definition. Let W be a linear manifold in X, and let $T(W)$ denote (as before) the algebra of all linear transformations of X with domain W and range contained in W. Given $T \in B(X)$ and $m = 0,1,\ldots,$ a $C^m[\alpha,\beta]$-operational calculus for T on W is an algebra homomorphism $\tau: C^m[\alpha,\beta] \to T(W)$ which sends the functions $\varphi(\lambda) \equiv 1$ and $\varphi(\lambda) \equiv \lambda$ to $I|W$ and $T|W$ respectively, and such that $\tau(\cdot)x: C^m[\alpha,\beta] \to X$ is continuous for each $x \in W$. Clearly, τ is unique (when it exists).

T is of class $C^m[\alpha,\beta]$ on W if there exists a $C^m[\alpha,\beta]$-operational calculus for T on W.

As in §4, we may (and shall) assume that $0 \in [\alpha,\beta]$. We shall say also that T is of class C^m on W if it is of class $C^m[\alpha,\beta]$ on W for some interval $[\alpha,\beta]$.

Note that if T is of class C^m on W, then W is necessarily T-invariant, since $T|W = \tau(\lambda) \in T(W)$. In general, if W is a T-invariant linear manifold in X, then $T(W)$ is a locally convex topological algebra for the topology induced by the semi-norms $T \to ||Tx||, x \in W$. The identity is $I|W$, and T is of class $C^m[\alpha,\beta]$ on W if and only if $T|W$ is of class $C^m[\alpha,\beta]$ as an element of the topological algebra $T(W)$.

14.2 Definition. Fix $T \in B(X)$ and an interval $[\alpha,\beta]$. Let $||f||_m$ denote the $C^m[\alpha,\beta]$-norm of f, and set

$$|x|_m = |x|_{m,T} = \sup\{||p(T)x||;\ p \in P,\ ||p||_m \leq 1\}$$

$(x \in X;\ m = 0,1,2,\ldots)$.

The C^m-manifold for T (over $[\alpha,\beta]$) is the set

$$W_m = \{x \in X;\ |x|_m < \infty\}.$$

Clearly, Lemma 13.14 is valid for $(W_m, |\cdot|_m)$ (instead of $(D_m, |\cdot|_m)$.

14.3 Theorem. Let W_m be the C^m-manifold for T (over $[\alpha,\beta]$). Then T is of class $C^m[\alpha,\beta]$ on W_m, and W_m is maximal with this property, for each

$m = 0,1,2,\ldots$. Moreover, there exists a unique continuous linear mapping U_m: $C[\alpha,\beta] \to T(W)$ (with $T(W)$ topologized as above) such that the $C^m[\alpha,\beta]$-operational calculus for T on W is given by

$$\tau_m(f) = \sum_{0 \le j \le m-1} \frac{f^{(j)}(0)}{j!} T^j + U_m(f^{(m)}) \quad (f \in C^m[\alpha,\beta]).$$

The theorem follows immediately from the density of P in $C^m[\alpha,\beta]$, the comments preceding Definition 14.2, and Theorem 4.1, formula (4.1.4).

14.4 We consider next the case of an unbounded operator T with real spectrum $\sigma(T)$. We shall discuss only the important special case $m = 0$.

Let $D(T)$ denote the domain of T. Since $\sigma(T) \subset R$, the function

$$R(t) = (I - itT)^{-1} \qquad (t \in R)$$

is a well-defined $B(X)$-valued function on R.

As in Section 4, let Ω denote the set of all vectors

$$(\underline{c};\underline{t}) = (c_1,\ldots,c_m;\ t_1,\ldots,t_m) \in C^m \times R^m$$

with m varying $(m = 1,2,\ldots)$, such that

$$\Big|\sum_{k=1}^{m} c_k \exp(it_k s)\Big| \le 1 \qquad (s \in R).$$

For any function F defined on R with values in a Banach space Y, we let

$$||F||_B = \sup_{\Omega} ||\textstyle\sum c_k F(t_k)||,$$

where $\sup_{\Omega}$ stands for the supremum over all vectors $(\underline{c};\underline{t})$ in Ω.

The set of all F: $R \to Y$ with $||F||_B < \infty$ is a vector space $F(Y)$ for the pointwise operations. The following elementary properties are easily verified:

(a) $F(Y)$ contains the constant functions $F(t) = c$ $(c \in Y)$, and $||c||_B = ||c||$.

(b) $||F||_B \ge ||F||_\infty = \sup_R ||F(t)||$.

(c) $(F(Y), ||\cdot||_B)$ is a Banach space.

(d) The norm $||\cdot||_B$ on $F(Y)$ is invariant under additive translations $F(t) \to F(t+c)$ $(c \in R)$ and multiplicative translations $F(t) \to F(ct)$ $(0 \ne c \in R)$.

(e) For any $U \in B(Y_1,Y_2)$ (where Y_1,Y_2 are Banach spaces), $||UF||_B \le ||U||\ ||F||_B$, so that

$$UF(Y_1) \subset F(Y_2).$$

(f) $F \in F(Y)$ if and only if $y^* F \in F(C)$ for all $y^* \in Y^*$;

in addition,

$$||F||_B = \sup\{||y^*F||_B;\ y^* \in Y^*,\ ||y^*|| \leq 1\}.$$

Also, if $F: R \to B(X,Y)$, then the following statements are equivalent

(i) $F \in F(B(X,Y))$;

(ii) $Fx \in F(Y)$ for each $x \in X$;

(iii) $y^*Fx \in F(C)$ for each $x \in X$ and $y^* \in Y^*$.

In addition,

$$||F||_B = \sup\{||Fx||_B;\ x \in X, ||x|| \leq 1\}$$

$$= \sup\{||y^*Fx||_B;\ x \in X, y^* \in Y^*, ||x|| \leq 1, ||y^*|| \leq 1\}.$$

Property (f) follows from the Uniform Boundedness theorem.

14.5 Definition. Let

$$|||x||| = \sup_{n \geq 0} ||R^n x||_B \qquad (x \in X)$$

where $R(t) = (I-itT)^{-1}$.

The semi-simplicity manifold for T is the set

$$Z = \{x \in X;\ |||x||| < \infty\}.$$

14.6 Lemma. (i) Z is a linear manifold in X, and $|||\cdot|||$ is a norm on Z which majorizes the given norm;

(ii) $(Z, |||\cdot|||)$ is a Banach space;

(iii) if $U \in B(X)$ commutes with T (that is, $UT \subset TU$), then $UZ \subset Z$ and $|||Ux||| \leq ||U||\ |||x|||$ for all $x \in Z$.

Proof. We show only the completeness of $(Z, |||\cdot|||)$. If $\{x_p\} \subset Z$ is $|||\cdot|||$-Cauchy, it is also $||\cdot||$-Cauchy. Let $x = \lim_{p \to \infty} x_p$ (in X). Given $\varepsilon > 0$, let p_0 be such that $|||x_p - x_q||| < \varepsilon$ for $p > q > p_0$. For each $(\underline{c}, \underline{t}) \in \Omega$ and $n \geq 0$,

$$||\sum c_k R^n(T_k)(x_p - x_q)|| < \varepsilon \quad (p > q > p_0).$$

Keeping q fixed and letting $p \to \infty$, we obtain

$$||\sum c_k R^n(t_k)(x - x_q)|| \leq \varepsilon \quad (q > p_0)$$

for all $(\underline{c},\underline{t}) \in \Omega$ and $n \geq 0$, that is

$$|||x-x_q||| \leq \varepsilon \qquad (q > p_0).$$

14.7 Lemma. The following statements are equivalent:

(i) $Z = X$;

(ii) Z is of the second category in X;

(iii) $|||\cdot|||$ is equivalent to $||\cdot||$;

(iv) $\sup_{n\geq 0} ||R^n||_B < \infty$.

Proof. Apply the Closed Graph theorem and the Uniform Boundedness theorem.

14.8 Definition. $T|Z$ is the restriction of T with domain

$$D(T|Z) = \{x \in D(T);\ x, Tx \in Z\} .$$

14.9 Lemma. For any $0 \neq t \in R$, $D(T|Z) = R(t)Z$.

Proof. If $x \in D(T|Z)$, then $z = (I-itT)x \in Z$ and therefore $x = R(t)z \in R(t)Z$ for any $t \in R$. Conversely, if $x \in R(t)Z$ for some $0 \neq t \in R$, say $x = R(t)z$ with $z = x-itTx \in Z$, then $x \in Z \cap D(T)$ (cf. Lemma 14.6(iii)) and $Tx = it^{-1}(z-x) \in Z$, that is $x \in D(T|Z)$.

14.10 Lemma. If iT generates a strongly continuous group of operators $\{G(t);\ t \in R\}$, then $|||x||| = ||Gx||_B$ $(x \in X)$.

Proof. By [11; Theorem 11.6.6],

$$G(t)x = \lim_{n\to\infty} R^n(t|n)x \qquad (t \in R, x \in X).$$

Therefore, for each $(\underline{c},\underline{t}) \in \Omega$ and $x \in X$,

$$\sum c_k G(t_k)x = \lim_{n\to\infty} \sum c_k R^n(t_k|n)x.$$

The sums on the right have norms bounded by

$$||R^n(t|n)x||_B = ||R^n x||_B \leq |||x|||$$

(cf. 14.4(d)). Hence $||Gx||_B \leq |||x|||$.

By [11, formula (5.8.5) and Theorem 11.5.2],

$$R(\lambda;\pm iT)^n x = \Gamma(n)^{-1}\int_0^\infty e^{-\lambda t} t^{n-1} G^{\pm}(t)x\,dt$$

for all $x \in X$, $n = 1,2,\ldots$, and $\lambda > 0$, where $G^{\pm}(t) = G(\pm t)$ $(t \geq 0)$ are the semigroups with generators $\pm iT$ respectively. Since

$$R^n(\pm 1/\lambda) = [\lambda R(\lambda;\pm iT)]^n,$$

a simple calculation leads to the formula

$$R^n(t)x = \Gamma(n)^{-1}\int_0^\infty e^{-s}s^{n-1}G(ts)x\,ds \qquad (14.10.1)$$

valid for all $x \in X$, $n = 1,2,\ldots,$ and $t \in \mathbb{R}$.

Since for each fixed $s > 0$

$$||G(ts)x||_B = ||Gx||_B \qquad \text{(cf. 14.4(d))},$$

therefore, if $||Gx||_B < \infty$ and $(\underline{c},\underline{t}) \in \Omega$,

$$||\sum c_k R^n(t_k)x|| = \Gamma(n)^{-1}||\int_0^\infty e^{-s}s^{n-1}\sum c_k G(t_k s)x\,ds||$$

$$\leq \Gamma(n)^{-1}\int_0^\infty e^{-s}s^{n-1}||G(ts)x||_B ds$$

$$= ||Gx||_B.$$

Hence $||R^n x||_B \leq ||Gx||_B$ for $n = 1,2,\ldots,$ and trivially $||R^0x||_B = ||x|| = ||G(0)x|| \leq ||Gx||_B$, that is, $|||x||| \leq ||Gx||_B$ is valid when $||Gx||_B < \infty$ and trivially otherwise, Q.E.D.

We now state our "local spectral theorem" for unbounded operators (cf. Definition 13.24 for the concept of a spectral measure on Z).

14.11 Theorem. Suppose iT is the infinitesimal generator of a strongly continuous group of operators in a reflexive Banach space X, and let Z be the semi-simplicity manifold for T. Then there exists a spectral measure on Z, E, which commutes with every $U \in B(X)$ commuting with T, such that

(i) $D(T|Z) = \{x \in Z;\ \lim_{n\to\infty}\int_{-n}^{n} sE(ds)x$ exists and belongs to $Z\}$, and

(ii) $Tx = \lim_{n\to\infty}\int_{-n}^{n} sE(ds)x, \quad x \in D(T|Z)$

(the limits are strong limits in X).

Proof. Let $\{G(t);\ t \in R\}$ be the group generated by iT. By Lemma 14.10, $||Gx||_B = |||x||| < \infty$ for each $x \in Z$.

For each $x \in Z$ and $x^* \in X^*$, the complex valued function $x^*G(\cdot)x$ is continuous on R and satisfies the "Bochner criterion" (cf. [2; 44])

$$\left|\sum_{k=1}^{m} c_k x^* G(t_k) x\right| \leq |||x|||\ ||x^*||\ \left|\left|\sum_{k=1}^{m} c_k \exp(it_k s)\right|\right|_\infty$$

for all vectors $(\underline{c},\underline{t}) \in C^m \times R^m$ $(m = 1,2,\ldots)$.

There exists therefore a **unique** regular complex Borel measure $\mu(\cdot|x,x^*)$ on $\mathcal{B}(R)$ such that

$$x^* G(t) x = \int_R e^{its} \mu(ds|x,x^*) \tag{14.11.1}$$

and

$$\operatorname{var} \mu(\cdot|x,x^*) \leq |||x|||\ ||x^*|| \tag{14.11.2}$$

for all $x \in Z$ and $x^* \in X^*$.

By reflexivity of X, we obtain a uniquely determined family of X-valued countably additive functions $\{E(\cdot)x|\ x \in Z\}$ on $\mathcal{B}(R)$ such that $\mu(\delta|x,x^*) = x^* E(\delta)x$ for all $x \in Z$, $x^* \in X^*$, and $\delta \in \mathcal{B}(R)$. For each $\delta \in \mathcal{B}(R)$, $E(\delta)$ is a linear operator with domain Z. By (14.11.1) with $t = 0$, $E(R) = I|Z$. We rewrite (14.11.1) in the "vector form"

$$G(t)x = \int_R e^{its} E(ds)x \qquad (t \in R;\ x \in Z). \tag{14.11.3}$$

If $U \in B(X)$ commutes with T (that is, $UT \subset TU$), then U commutes with $G(t)$ for all $t \in R$, and $Ux \in Z$ for $x \in Z$ (cf. Lemma 14.6 (iii)). It then follows from the uniqueness of the representation (14.11.3) that $UE(\delta)x = E(\delta)Ux$ for all $x \in Z$ (i.e, U commutes with $E(\delta)$).

For $h \in \underline{B}(R)$ (the Banach algebra of all bounded complex Borel functions on R with the supremum norm), define

$$\Phi(h)x = \int_R h(s)E(ds)x \qquad (x \in Z).$$

Since $E(\cdot)x$ is (strongly) countably additive (for $x \in Z$), $\Phi(h)$ is a well-defined linear operator in X with domain Z, and by (14.11.2),

$$||\Phi(h)x|| \leq ||h||_\infty\ |||x||| \qquad (h \in \underline{B}(R),\ x \in Z).$$

In addition, $\Phi(h)$ commutes with every $U \in B(X)$ which commutes with T.

We now improve the above inequality, replacing the $||\cdot||$-norm by the larger $|||\cdot|||$-norm on the left side. Fix $x \in Z$ and $(\underline{c},\underline{t}) \in \Omega$. For each

$(\underline{c}',\underline{t}') \in \Omega$,

$$||\sum_j c_j' G(t_j') \sum_k c_k G(t_k)x|| = ||\sum_{j,k} c_j' c_k G(t_j'+t_k)x||$$

$$\leq |||x|||\ ||\sum_{j,k} c_j' c_k \exp i(t_j'+t_k)s||_\infty$$

$$\leq |||x|||\ ||\sum_j c_j' \exp it_j's||_\infty ||\sum_k c_k \exp it_k s||_\infty \leq |||x|||.$$

Hence, by Lemma 14.10,

$$|||\sum_k c_k G(t_k)x||| \leq |||x||| \quad (x \in Z, (\underline{c},\underline{t}) \in \Omega).$$

Fix $h \in \underline{B}(R)$ and $x \in Z$. For each $(\underline{c},\underline{t}) \in \Omega$,

$$||\sum_k c_k G(t_k)\Phi(h)x|| = ||\Phi(h)\sum_k c_k G(t_k)x||$$

$$\leq ||h||_\infty |||\sum_k c_k G(t_k)x||| \leq ||h||_\infty |||x|||$$

(since $\sum c_k G(t_k)$ commutes with T). Taking the supremum over all $(\underline{c},\underline{t}) \in \Omega$, we obtain

$$|||\Phi(h)x||| \leq ||h||_\infty |||x||| \qquad (h \in \underline{B}(R),\ x \in Z). \qquad (14.11.4)$$

In particular,

$$\Phi(h)Z \subset Z \qquad (h \in \underline{B}(R)).$$

Taking $h = \chi_\delta$, we see that $|||E(\delta)x||| \leq |||x|||$ (i.e., $E(\delta) \in T(Z)$) for all $\delta \in \mathcal{B}(R)$ and $x \in Z$. Since for all parameters involved (a.p.i)

$$\int_R e^{ius}\mu(ds|G(t)x,x^*) = x^*G(u)G(t)x = x^*G(u+t)x$$

$$= \int_R e^{ius}e^{its}\mu(ds|x,x^*),$$

it follows, by regularity of μ and the uniqueness theorem for Fourier-Stieltjes transforms, that $\mu(\delta|G(t)x,x^*) = \int_R e^{its}\chi_\delta(x)\mu(ds|x,x^*)$ (a.p.i.).
However the left side is equal to

$$x^*E(\delta)G(t)x = x^*G(t)E(\delta)x$$

$$= \int_R e^{its}\mu(ds|E(\delta)x,x^*)$$

since $E(\delta)x \in Z$ for $x \in Z$.

Hence, again by uniqueness for Fourier-Stieltjes transforms,

$$x^*E(\sigma)E(\delta)x = \mu(\sigma|E(\delta)x,x^*)$$
$$= \int_R \chi_\sigma(x)\chi_\delta(x)\mu(ds|x,x^*)$$
$$= \mu(\sigma\cap\delta|x,x^*) = x^*E(\sigma\cap\delta)x$$

for all $x \in Z$, $x^* \in X^*$, and $\sigma,\delta \in \mathcal{B}(R)$, that is, $E(\sigma)E(\delta) = E(\sigma\cap\delta)$ $(\sigma,\delta \in \mathcal{B}(R))$, and we conclude that <u>E is a spectral measure on Z</u>. Equivalently,

$$E(ds)E(\delta)x = \chi_\delta(s)E(ds)x \qquad (\delta \in \mathcal{B}(R),\ x \in Z).$$

Hence $\phi(h\chi_\delta) = \phi(h)\phi(\chi_\delta)$ for all $h \in \underline{B}(R)$ and $\delta \in \mathcal{B}(R)$, and so $\phi(hg) = \phi(h)\phi(g)$ for all $h \in \underline{B}(R)$ and $g \in \underline{B}_0(R)$ (the subalgebra of simple functions in $\underline{B}(R)$). For $h,g \in \underline{B}(R)$, choose $g_n \in \underline{B}_0(R)$ such that $||g_n-g||_\infty \to 0$. By (14.11.4), for all $x \in Z$,

$$||\phi(hg)x - \phi(h)\phi(g)x|| \le ||\phi\,[h(g-g_n)]x|| + ||\phi(h)\phi(g_n-g)x||$$
$$\le ||h(g-g_n)||_\infty\, |||x||| + ||h||_\infty |||\phi(g_n-g)x|||$$
$$\le 2||h||_\infty ||g_n-g||_\infty\, |||x||| \to 0 \quad \text{as} \quad n \to \infty.$$

We conclude that Φ: $\underline{B}(R) \to T(Z)$ is an <u>algebra homomorphism</u>, continuous with respect to the $B(Z,|||\cdot|||)$-topology on $T(Z)$.

By (14.10.1) and (14.11.3), we have for $x \in Z$ and $t \in R$,

$$x^*R(t)x = x^*\int_0^\infty e^{-u}G(tu)x\,du$$
$$= \int_0^\infty e^{-u}\int_R e^{itus}x^*E(ds)x\,du.$$

Since $\int_0^\infty\int_R e^{-u}|x^*E(ds)x|du = \mathrm{var}(x^*Ex) \le ||x^*||\ |||x|||$ by (14.11.2), Fubini's theorem implies that

$$x^*R(t)x = \int_R\int_0^\infty e^{-u}e^{itus}du\ x^*E(ds)x$$
$$= \int_R (1-ist)^{-1}\ x^*E(ds)x,$$

that is

$$R(t)x = \int_R (1-ist)^{-1}E(ds)x \qquad (14.11.5)$$

for all $t \in R$ and $x \in Z$.

Now let $x \in D(T|Z)$. By Lemma 14.9, $x = R(t)z$ with $z \in Z$ and $t \neq 0$ fixed. By multiplicativity of Φ and (14.11.5),

$$\int_{-n}^{n} sE(ds)x = \int_{-n}^{n} s(1-ist)^{-1}E(ds)z \to \int_{R} s(1-ist)^{-1}E(ds)z.$$

Writing $s(1-ist)^{-1} = it^{-1}[1-(1-ist)^{-1}]$, the last integral is seen to equal

$$it^{-1}[z-R(t)z] = it^{-1}(z-x) = Tx \in Z.$$

Thus the inclusion $\subset$ in Theorem 14.11(i) and Relation(ii) are proved.

On the other hand, if x belongs to the set on the right hand side of (i), then denoting the limit in (i) by $z \in Z$, we have by (14.11.5) and the multiplicativity of Φ (for $t \neq 0$):

$$\begin{aligned} R(t)z &= \lim_{n\to\infty} \int_{-n}^{n} sE(ds)R(t)x \\ &= \lim_{n\to\infty} \int_{-n}^{n} s(1-its)^{-1}E(ds)x \\ &= \int_{R} s(1-its)^{-1}E(ds)x = it^{-1}[x-R(t)x] \end{aligned}$$

(cf. preceding calculation). Therefore

$$x = R(t)(x-itz) \in R(t)Z = D(T|Z)$$

by Lemma 14.9, and the proof of Theorem 14.11 is complete.

14.12 <u>The case $Z = X$.</u>

By Lemma 14.7, we have $Z = X$ if and only if

$$V_T = \sup_{n\geq 0} ||R^n||_B < \infty.$$

14.13 <u>Lemma</u>. Let T be a possibly unbounded operator in a Banach space X. Then T is closed with dense domain, real spectrum, and $V_T < \infty$ if and only if iT generates a strongly continuous group of operators $\{G(t);\ T \in R\}$ with $||G||_B < \infty$.

In this case, $V_T = ||G||_B$.

<u>Proof</u>. 1. If T has real spectrum, then V_T is well-defined; assume $V_T < \infty$. Then, for $0 \neq \lambda \in R$ and $n = 1,2,\ldots$,

$$||[\lambda R(\lambda;iT)]^n|| = ||R^n(\lambda^{-1})|| \leq ||R^n||_B \leq V_T < \infty.$$

Therefore, if T is also closed with dense domain, the Hille-Yosida theorem (cf. [11; Theorem 12.3.2]) implies that iT generates a strongly continuous group of operators $\{G(t);\ t \in R\}$.

By Lemma 14.10, $||Gx||_B = |||x||| \leq V_T||x||$ for all x, and therefore $||G||_B \leq V_T$ (cf. 14.4(f)).

2. Conversely, if iT generates a strongly continuous group of operators G with $||G||_B < \infty$, then since $||G||_\infty \leq ||G||_B$, we have

$$\lim_{t\to\infty} t^{-1} \log||G^{\pm}(t)|| \leq 0,$$

where as before $G^{\pm}(t) = G(\pm t)$ $(t \geq 0)$ are the semi-groups with generators $\pm iT$ respectively. By [11; p.622], $\sigma(\pm iT)$ is contained in the closed left half-plane. Hence $\sigma(iT)$ is pure imaginary and $\sigma(T) \subset R$. As a (semi)group generator, iT is closed and densely defined.

By Lemma 14.10,

$$||R^n x||_B \leq |||x||| = ||Gx||_B \leq ||G||_B||x|| \qquad (n \geq 0)$$

for all $x \in X$, hence $||R^n||_B \leq ||G||_B$ for all $n \geq 0$ (cf. 14.4(f)), and so $V_T \leq ||G||_B$. Q.E.D.

We now recall that an unbounded operator T with real spectrum is said to be <u>scalar-type spectral</u> if there exists a (strongly) countably additive spectral measure E on $\mathcal{B}(R)$ such that

(i) $D(T) = \{x \in X;\ \lim_{n\to\infty} \int_{-n}^{n} sE(ds)x \text{ exists}\}$

and

(ii) $Tx = \lim_{n\to\infty} \int_{-n}^{n} sE(ds)x \qquad (x \in D(T))$.

The following result is an easy corollary of Theorem 14.11.

14.14 <u>Theorem</u>. Let T be a possibly unbounded linear operator with real spectrum, acting in a reflexive Banach space. Then T is a scalar-type spectral operator if and only if it is closed, densely defined, and $V_T < \infty$.

<u>Proof</u>. 1. <u>Sufficiency</u>. By Lemma 14.13, iT generates a strongly continuous group of operators G with $||G||_B = V_T < \infty$. Also $Z = X$ and the norms $|||\cdot|||$ and $||\cdot||$ are equivalent, by Lemma 14.7. Let E be the spectral measure on X $(= Z)$ provided by Theorem 14.11. Since $|||E(\delta)x||| \leq |||x|||$ for all $x \in X$ and $\delta \in \mathcal{B}(R)$, the equivalence of the norms implies that $E: \mathcal{B}(R) \to B(X)$ is a spectral measure in the usual sense. Properties (i) and (ii) in Theorem 14.11 trivially reduce to the defining properties of scalar-type spectrality.

2. Necessity (valid without the reflexivity hypothesis). Let T be a scalar-type spectral operator, and let E be the resolution of the identity for T. The fact that T is closed and densely defined is well-known. Next

$$R^n(t) = \int_R (1-its)^{-n} E(ds) \qquad (t \in R, n \geq 0).$$

Let $f_n(u) = \Gamma(n)^{-1} u^{n-1} e^{-u}$ $\qquad (u \geq 0;\ n = 1,2,\ldots)$.

Then

$$\int_0^\infty e^{ivu} f_n(u)\,du = (1-iv)^{-n} \qquad (v \in R).$$

Hence, for all $(\underline{c},\underline{t}) \in \Omega$ and $n \geq 1$,

$$\sum c_k R^n(t_k) = \int_R \int_0^\infty \sum c_k \exp(it_k su) \cdot f_n(u)\,du E(ds).$$

The sum appearing in the integrand is bounded by 1; $f_n \geq 0$ and $\int_0^\infty f_n(u)\,du = 1$.

It follows that the expression above has norm $\leq v(E)$, where $v(E)$ denotes the norm of the map $h \to \int_R h(s)E(ds)$ of $\underline{B}(R)$ into $B(X)$. Hence $V_T \leq v(E) < \infty$.

14.15 Spectrum in a half-line.

The hypothesis that iT generates a group, which appears in Theorem 14.11, may be superfluous, and was dropped indeed in the extreme case $Z = X$. However, our proof does depend on this hypothesis. A way out of this difficulty is to work in the $|||\cdot|||$-closure of $D(T|Z)$, and appeal to [11; Theorem 12.2.4] and to the Hille-Yosida theorem in that space. However the latter is not necessarily reflexive (since $|||\cdot|||$ is larger than the given X-norm), and the construction of the spectral measure on Z cannot proceed in the usual way.

In case $\sigma(T)$ is contained in a half-line ($[0,\infty)$ without loss of generality), a local spectral theorem can be proved without the group generation hypothesis. The construction of Z will proceed however along different lines.

14.16 Notation. Let M and D denote respectively the formal operators of multiplication

$$M: f(t) \to tf(t)$$

and differentiation

$$D: f(t) \to f'(t).$$

The Widder (formal) differential operators L_k are given by

$$L_k = c_k M^{k-1} D^{2k-1} M^k \qquad k = 1,2,\ldots$$

where $c_1 = 1$ and $c_k = (-1)^{k-1}[\Gamma(k-1)\Gamma(k+1)]^{-1}$ for $k \geq 2$.

For $s,t > 0$, $B(x,t)$ will denote the Beta function

$$B(s,t) = \Gamma(s)\Gamma(t)\Gamma(s+t)^{-1}.$$

We use $||\cdot||_1$ for the $L^1(R^+, dt/t)$-norm.

Suppose (throughout) that $\sigma(T) \subset [0,\infty)$, and let then R be the resolvent of $-T$, that is $R(t) = (t+T)^{-1}$, for $t \in R^+$. We set

$$S = TR(I-TR).$$

This is a well-defined $B(X)$-valued function on R^+. Since

$$S(t) = tR(t)[I-tR(t)] \quad (t > 0),$$

the functions S^k $(k = 1,2,\ldots)$ are of class C^∞ on R^+.

For $k = 0,1,2,\ldots$, we set

$$||x||_k = ||x|| \qquad k = 0$$

$$= \sup\{B(k,k)^{-1}||x^*S^kx||_1;\ x^* \in X^*, ||x^*|| \leq 1\} \qquad k \geq 1,$$

and $$|||x||| = \sup_{k\geq 0} ||x||_k.$$

In the present context, the <u>semi-simplicity manifold</u> for T is the set

$$Z = \{x \in X;\ |||x||| < \infty\}.$$

14.17 <u>Lemma</u>. 14.6 is valid in the present context.

<u>Proof</u>. We verify only the completeness of $(Z, |||\cdot|||)$. Let $\{x_n\} \subset Z$ be $|||\cdot|||$-Cauchy. since $|||\cdot||| \geq ||\cdot||$, $\{x_n\}$ converges to some $x \in X$ in the $||\cdot||$-norm. For each $k \geq 0$ and $x^* \in X^*$ with $||x^*|| \leq 1$, $x^*S^k(\cdot)x_n \to x^*S^k(\cdot)x$ pointwise as $n \to \infty$, and

$$B(k,k)^{-1}||x^*S^kx_n||_1 \leq |||x_n||| \leq K \quad \text{for all } n.$$

By Fatou's lemma, it follows that $|||x||| \leq K$, i.e., $x \in Z$. Now, given $\varepsilon > 0$, let n_0 be such that $|||x_n - x_m||| < \varepsilon$ for $n > m > n_0$. Since $x^*S^k(x-x_m) = \lim_{n\to\infty} x^*S^k(x_n-x_m)$ pointwise, we have again by Fatou's lemma,

$$B(k,k)^{-1}\|x^*S^k(x-x_m)\|_1 \leq \varepsilon$$

for all $k \geq 1$, $x^* \in X^*$ with $\|x^*\| \leq 1$, and $m > n_0$.

Hence $|||x-x_m||| \leq \varepsilon$ for $m > n_0$, Q.E.D.

The notation $D(T|Z)$ is the same as in Definition 14.8 (with the new Z!). As before, we have

$$D(T|Z) = R(t)Z$$

for any $t > 0$ (cf. Lemma 14.9's proof).

We state now the local spectral theorem for operators with spectrum in a half-line (or more generally, with $(-\infty,0) \subset \rho(T)$.

14.13 <u>Theorem</u>. Let T be a possibly unbounded linear operator in the reflexive Banach space X, with spectrum in $[0,\infty)$. Let Z be the semi-simplicity manifold for T. Then there exists a spectral measure on Z, E, which commutes with every $U \in B(X)$ commuting with T, such that

(i) $D(T|Z) = \{x \in Z;\ \lim_{n\to\infty} \int_0^n sE(ds)x$ exists and belongs to $Z\}$, and

(ii) $Tx = \lim_{n\to\infty} \int_0^n sE(ds)x, \qquad x \in D(T|Z)$

(the limits are strong limits in X).

<u>Proof</u>. Let L_k $(k = 1,2,\dots)$ be the Widder formal differential opeators (cf. Section 14.16). By Leibnitz' rule,

$$L_k = c_k' \sum_{j=0}^{k} \Gamma(k+j)^{-1}\binom{k}{j} M^{k+j-1}D^{k+j-1},$$

where $c_1' = 1$ and $c_k' = (-1)^{k-1}B(k-1,k+1)$ for $k \geq 2$.

For $x \in X$ and $x^* \in X^*$ fixed,

$$D^{k+j-1}(x^*Rx) = (-1)^{k+j-1}\Gamma(k+j)\,x^*R^{k+j}x,$$

and therefore

$$L_k x^*R(t)x = c_k'' t^{-1} x^*(tR)^k \sum_{j=0}^{k}\binom{k}{j}(-tR)^j x$$
$$= c_k'' t^{-1} x^*S^k(t)x,$$

where $c_1'' = 1$ and $c_k'' = B(k-1,k+1)^{-1}$ for $k \geq 2$.

Hence

$$\int_0^\infty |L_k(x^*Rx)|dt = c_k''||x^*S^kx||_1. \qquad (14.18.1)$$

If $x \in Z$, we have

$$||x^*S^kx||_1 \leq B(k,k)\,|||x|||\;||x^*|| \qquad (x^* \in X^*;\ k \geq 1) \qquad (14.18.2)$$

and therefore

$$\int_0^\infty |L_k(x^*Rx)|dt \leq |||x|||\;||x^*|| \qquad (14.18.3)$$

for all $k \geq 1$, $x \in Z$, and $x^* \in X^*$ (since $B(k,k)/B(k-1,k+1) = \frac{k-1}{k} < 1$ for $k \geq 2$).

We now rely on the following complex version of the Widder STR (Stieltjes Transform Representation) Theorem (cf. [48; Theorem 16, p.361]):

Let f be a C^∞ complex function on R^+ such that

$$K = \sup_{k\geq 1}\int_0^\infty |L_kf|dt < \infty.$$

then the limit $A = \lim_{t\to 0+} tf(t)$ exists, and there is a unique complex regular Borel measure μ on R^+ such that

$$\text{var}\,\mu \leq 2K+|A|$$

and

$$f(t) = \int_0^\infty (t+s)^{-1}\mu(ds) \qquad (t \in R^+).$$

Taking $f(t) = x^*R(t)x$ $(x^* \in X^*, x \in Z)$, we have $K \leq |||x|||\;||x^*||$ by (14.18.3). Since $\lim_{t\to 0+} x^*tR(t)x$ exists for each $x^* \in X^*$ (by the STR theorem), the Uniform Boundedness theorem implies that

$$H_x = \sup_{t>0}||tR(t)x|| < \infty \qquad (x \in Z). \qquad (14.18.4)$$

Let $M_x = 2|||x||| + H_x$. By the STR theorem, there exists for each $x \in Z$ and $x^* \in X^*$ a unique complex regular Borel measure $\mu(\cdot|x,x^*)$ such that

$$\text{var}\,\mu(\cdot|x,x^*) \leq M_x||x^*|| \qquad (x \in Z, x^* \in X^*) \qquad (14.18.5)$$

and

$$x^*R(t)x = \int_0^\infty (t+s)^{-1}\mu(ds|x,x^*) \qquad (t \in R^+). \qquad (14.18.6)$$

The uniqueness of the representation and (14.18.5) imply that $\mu(\delta|x,\cdot)$ is a continuous linear functional on X^*, for each fixed $\delta \in \mathcal{B}(R^+)$ and $x \in Z$.

Since X is reflexive, there exists a unique function $E(\cdot)x: \mathcal{B}(R^+) \to X$ (for each fixed $x \in Z$) such that $\mu(\delta|x,x^*) = x^*E(\delta)x$ for all $\delta \in \mathcal{B}(R^+)$ and $x^* \in X^*$. Necessarily, $E(\delta)$ is a linear operator with domain Z, and

$$||E(\delta)x|| \leq M_x \quad (\delta \in \mathcal{B}(R^+);\ x \in Z). \tag{14.18.5'}$$

For $x \in Z$ fixed, $E(\cdot)x$ is weakly, hence strongly countably additive, and $R(t)x = \int_0^\infty (t+s)^{-1}E(ds)x \quad (x \in Z, t > 0)$.

If $U \in B(X)$ commutes wtih T, then U commutes with $R(t)$ for all $t > 0$, and $UZ \subset Z$.

Hence
$$\int_0^\infty (t+s)^{-1}E(ds)Ux = R(t)Ux = UR(t)x$$
$$= \int_0^\infty (t+s)^{-1}UE(ds)x \qquad (x \in Z, t > 0)$$

and so $E(\delta)Ux = UE(\delta)x$ for all $x \in Z$ and $\delta \in \mathcal{B}(R^+)$, by the uniqueness claim in the STR theorem.

Taking in particular $U = R(u)$ $(u > 0$ fixed), we obtain for $x \in Z$ (using the integral representation of $R(t)[R(u)x]$)

$$R(u)E(R^+)x = E(R^+)R(u)x$$
$$= \lim_{t\to\infty} \int_0^\infty t(t+s)^{-1}E(ds)R(u)x$$
$$= \lim_{t\to\infty} tR(t)R(u)x$$
$$= \lim_{t\to\infty}\{t(t-u)^{-1}R(u)x - tR(t)x/(t-u)\}$$
$$= R(u)x,$$

by the first resolvent equation and (14.18.4). Since $R(u)$ is one-to-one, we conclude that $E(R^+) = I|Z$.

For $t,u > 0$, $t \neq u$, and $x \in Z$, we have by the first resolvent equation (and the fact $R(u)x \in Z$):

$$\int_0^\infty (t+s)^{-1}E(ds)R(u)x = R(t)R(u)x$$
$$= (t-u)^{-1} \int_0^\infty [(u+s)^{-1} - (t+s)^{-1}]E(ds)x$$
$$= \int_0^\infty (t+s)^{-1}(u+s)^{-1}E(ds)x.$$

By uniqueness of the STR,

$$E(\delta)R(u)x = \int_0^\infty (u+s)^{-1}\chi_\delta(s)E(ds)x \qquad (14.18.7)$$

for all $u > 0, \delta \in \mathcal{B}(R^+)$, and $x \in Z$.

Briefly

$$E(ds)R(u)x = (u+s)^{-1}E(ds)x, \qquad (14.18.8)$$

and inductively, for $k = 0,1,2,\ldots,$

$$E(ds)R(u)^k x = (u+s)^{-k}E(ds)x.$$

Hence

$$E(ds)p(R(u))x = p((u+s)^{-1})E(ds)x$$

for all polynomials p.

In particular,

$$E(ds)S^k(u)x = u^k(u+s)^{-k}[1-u(u+s)^{-1}]^k E(ds)x$$

$$= (us)^k(u+s)^{-2k}E(ds)x$$

for all $u > 0$, $x \in Z$, and $k = 0,1,2,\ldots$.

Since $S^k(u)$ commutes with $R(t)$, it commutes with $E(\delta)$, and therefore

$$x^*S^k(u)E(\delta)x = \int_\delta (us)^k(u+s)^{-2k}x^*E(ds)x.$$

By Tonelli's theorem and (14.18.5),

$$||x^*S^kE(\delta)x||_1 \le \int_\delta \int_0^\infty (us)^k(u+s)^{-2k}(du/u)\,|x^*E(ds)x|$$

$$= \int_\delta \int_0^\infty t^k(1+t)^{-2k}(dt/t)\ \ x^*E(ds)x|$$

$$= B(k,k)\,|x^*E(\delta)x|$$

$$\le B(k,k)M_x||x^*||$$

for all $x^* \in X^*$, $k \ge 1$, $\delta \in \mathcal{B}(R^+)$, and $x \in Z$.

Hence

$$|||E(\delta)x||| \le M_x \quad (x \in Z,\ \delta \in \mathcal{B}(R^+)).$$

In particular, $E(\delta) \in T(Z)$ $(\delta \in \mathcal{B}(R^+))$. Now, for each $x \in Z$, since $E(\delta)x \in Z$, we have

$$R(u)E(\delta)x = \int_0^\infty (u+s)^{-1}E(ds)E(\delta)x \qquad (u > 0).$$

Since $R(u)$ commutes with $E(\delta)$, it follows from (14.18.7) and the uniqueness of the STR that

$$E(\sigma)E(\delta)x = \int_0^\infty \chi_\sigma(s)\chi_\delta(s)E(ds)x = E(\sigma\cap\delta)x$$

for all $\sigma,\delta \in \mathcal{B}(R^+)$ and $x \in Z$.

In conclusion, E is a spectral measure on Z.

We verify finally Statements (i) and (ii) of the theorem. First, observe that $D(T|Z) = R(t)Z$ for any $t > 0$ (cf. Lemma 14.9). Thus, if $x \in D(T|Z)$, then $x = R(t)y$ for $t > 0$ and suitable $y \in Z$. By (14.18.8),

$$\int_0^n sE(ds)x = \int_0^n s(t+s)^{-1}E(ds)y$$

$$\underset{n\to\infty}{\to} \int_0^\infty s(t+s)^{-1}E(ds)y$$

$$= \int_0 [1-t(t+s)^{-1}]E(ds)y = [I-tR(t)]y \ (\in Z)$$

$$= TR(t)y = Tx.$$

If Z_1 denotes the set on the right hand side of (i), we have proved that $D(T|Z) \subset Z_1$ and that (ii) is valid. On the other hand, if $x \in Z_1$, denote the limit in (i) by $z \in Z$. For any $t > 0$, we have by (14.18.8) and the commutativity of E and $R(t)$

$$R(t)z = \lim_{n\to\infty} \int_0^n sR(t)E(ds)x$$

$$= \lim_{n\to\infty} \int_0^n s(t+s)^{-1}E(ds)x = \int_0^\infty s(t+s)^{-1}E(ds)x$$

$$= x-tR(t)x.$$

Hence $x = R(t)(z+tx) \in R(t)Z = D(T|Z)$.

Therefore $D(T|Z) = Z_1$, and the proof is complete.

As before, the important special case $Z = X$ gives the following result.

14.19 <u>Theorem</u>. Let T be a possibly unbounded operator in a reflexive Banach space X, with $\sigma(T) \subset [0,\infty)$. Then the following statements are equivalent

(a) $Z = X$.

(b) $K = \sup_{||x|| \le 1} |||x||| < \infty$.

(c) T is spectral of scalar type.

Proof. Since $|||x||| \ge ||x||$ for all x, the equivalence of (a) and (b) follows from the Closed Graph theorem. We show that (b) $\Rightarrow$ (c). By (14.18.4) and the Uniform Boundedness theorem,

$$H = \sup_{t>0} ||tR(t)|| < \infty,$$

and so

$$M_x \le M||x|| \quad (x \in X),$$

where $M = 2K+H$. Hence, by (14.18.5'), $E(\delta) \in B(X)$ (in fact, $||E(\delta)|| \le M$) for each $\delta \in B(R^+)$. Thus E is a spectral measure in the usual sense, and Properties (i) and (ii) in Theorem 14.18 reduce to the defining properties of scalar-type spectrality of the operator T (since $\sigma(T) \subset [0,\infty)$).

(c) $\Rightarrow$ (a) (no reflexivity needed).

Since $TR(t) = I - tR(t)$ $(t > 0)$, therefore

$$S(t) = t[I-tR(t)]R(t) = tTR(t)^2 \quad (t > 0).$$

Let E be the resolution of the identity for T. Then, for $t > 0$ and $k = 1,2,\ldots$,

$$\begin{aligned} S^k(t) &= t^k T^k (t+T)^{-2k} \\ &= t^k \int_0^\infty s^k (t+s)^{-2k} E(ds) \\ &= \int_0^\infty (s/t)^k (1+s/t)^{-2k} E(ds). \end{aligned}$$

For each $x \in X$ and $x^* \in X^*$,

$$||x^* S^k x||_1 \le \int_0^\infty \int_0^\infty (s/t)^k (1+s/t)^{-2k} |x^* E(ds) x| \, dt/t.$$

If we interchange the order of integration and substitute $u = s/t$ in the inner integral, the right hand side becomes

$$\int_0^\infty \int_0^\infty u^k (1+u)^{-2k} \, du/u \, |x^* E(ds) x| = B(k,k) \operatorname{var}(x^* E x) \le B(k,k)(\operatorname{var} E) ||x|| \, ||x^*||.$$

Applying Tonelli's theorem, we conclude that

$$|||x||| \le (\operatorname{var} E) ||x|| \quad (x \in X)$$

and (a) follows.

14.20 Local semigroups.

Theorem 14.11 may be interpreted as a spectral representation theorem for groups of operators. A refinement of its proof leads to a result on the spectral representation of local semigroups as well.

Let $\mathcal{D}$ be a linear manifold in the Banach space X. Let $\Delta = [0,a)$, where $0 < a \leq \infty$, and let $P: \Delta \to T(\mathcal{D})$ satisfy the following requirements:

(i) $P(0) = I|\mathcal{D}$.

(ii) $P(t+s) = P(t)P(s) \quad (t,s,t+s \in \Delta)$.

(iii) $P(\cdot)x \in C(\Delta,X) \quad (x \in \mathcal{D})$,

where $C(\Delta,X)$ denotes the space of X-valued continuous functions on Δ. The triple $(P,\mathcal{D},\Delta)$ (or shortly, P) is called a local semigroup with (common) domain $\mathcal{D}$ on Δ.

An important class of local semigroups arises as follows. Let E be a spectral measure on R, and let T be the scalar-type spectral (s.t.s.) operator whose resolution of the identity is E. Define $P(t) = e^{-tT}$ by means of the operational calculus for s.t.s. operators. Thus, for $t > 0$,

$$D(P(t)) = \{x \in X \mid \lim_{n\to\infty} \int_{\{s\in R \mid e^{-ts}\leq n\}} e^{-ts}E(ds)x \text{ exists}\}$$

$$= \{x \in X \mid \lim_{\omega\to-\infty} \int_{\omega}^{\infty} e^{-ts}E(ds)x \text{ exists}\}.$$

For all $t \in \Delta = [0,\infty)$ and $x \in D(P(t))$, we set

$$P(t)x = \lim_{\omega\to-\infty} \int_{\omega}^{\infty} e^{-ts}E(ds)x \; (= e^{-tT}x).$$

Let

$$\mathcal{D}_0 = \bigcup_{n=1}^{\infty} E([-n,n])X.$$

Clearly $\mathcal{D}_0 \subset \bigcap_{t\in\Delta} D(P(t))$, and one sees without difficulty that $(P,\mathcal{D}_0,\Delta)$ is a local semigroup. The same is true of $(P,\mathcal{D},\Delta)$ for any linear manifold $\mathcal{D} \subsetneq \mathcal{D}_0$ invariant for all $P(t)$ $(t \in \Delta)$. Any $(P,\mathcal{D},\Delta)$ of this type will be called a spectral local semigroup. Note that the local semigroup $(P,\mathcal{D}_0,\Delta)$ is "densely defined", by which we mean that the common domain $\mathcal{D}_0$ is dense in X.

14.21 Binomial vectors.

Let $\mathcal{P} = (P,\mathcal{D},\Delta)$ be an arbitrary local semigroup. Fix a rational number $c \in \Delta$, $c = p/q > 0$ $(p,q \in N)$. We say that $x \in \mathcal{D}$ is a binomial vector for $\mathcal{P}$ if there exists $k(x) \in N$ such that

$$P_k(z)x = \sum_{n=0}^{\infty} \binom{z}{n}[P(c/k)-I]^n x$$

is an entire function of z for all $k \geq k(x)$. The set of all binomial vectors for $\mathcal{P}$ is clearly a linear manifold $\mathcal{b} \subset \mathcal{D}$.

With Ω and $||\cdot||_B$ as in §14.4, we set

$$|||x||| = \sup_{k \geq k(x)} ||P_k(it)x||_B \qquad (x \in \mathcal{b}),$$

and

$$Z = \{x \in \mathcal{b};\ |||x||| < \infty\}.$$

For any $x \in \mathcal{b}$,

$$||P_k(it)x||_B \geq ||P_k(it)x||_\infty \geq ||P_k(0)x|| = ||x|| \quad (k \geq k(x)),$$

so that $|||x||| \geq ||x||$.

Z is a linear manifold $(Z \subset \mathcal{b})$ and $|||\cdot|||$ is a norm on Z.

14.22 Lemma. Let $x \in Z$. Then for $k \geq k(x)$, $P_{pk}(qkz)x$ is independent of k for all $z \in C$. Defining

$$\tilde{P}(z)x = P_{pk}(qkz)x \qquad (z \in C)$$

where $k \geq k(x)$, the function $\tilde{P}(.)x$ is entire and coincides with $P(.)x$ on Δ. Moreover, there exists a uniquely determined family $\{\mu(\cdot|x,x^*); x^* \in X^*\}$ of regular complex Borel measures with compact support on R such that

$$\operatorname{var} \mu(\cdot|x,x^*) \leq |||x|||\ ||x^*|| \qquad (x^* \in X^*)$$

and

$$x^*\tilde{P}(z)x = \int_R e^{-zv}\mu(dv|x,x^*) \qquad (z \in C,\ x^* \in X^*).$$

Proof. Let $x \in Z$ and $x^* \in X^*$. The function $x^*P_k(it)x$ (for $k \geq k(x)$) is continuous and satisfies the Bochner criterion (cf. [2]). Therefore there exists a unique regular complex Borel measure $\mu_k(\cdot|x,x^*)$ on R such that

$$\operatorname{var} \mu_k(\cdot|x,x^*) \leq |||x|||\ ||x^*|| \qquad (14.22.1)$$

and

$$x^*P_k(is)x = \int_R e^{-isv}\,\mu_k(dv|x,x^*) \qquad (14.22.2)$$

for all $x^* \in X^*$, $k \geq k(x)$, and $s \in R$.

Consider the bounded functions $x^*P_k(i\cdot)x$ as distributions on the Schwartz space $\mathcal{S}(R)$. Since they extend to the entire functions $x^*P_k(z)x$, and the latter are of exponential type by a well-known property of binomial series (cf. [11; p.233]), the Paley-Wiener-Schwartz theorem [10; Theorem 5, p. 145] implies that their Fourier transforms (which are $2\pi\mu_k(\cdot|x,x^*)$ by (14.22.2)) have compact support. Therefore the bilateral Laplace transform $\int_R e^{-zv}\mu_k(dv|x,x^*)$ is well-defined, entire, and coincides with $x^*P_k(z)x$ on the imaginary axis (by (14.22.2)). Hence

$$x^*P_k(z)x = \int_R e^{-zv}\mu_k(dv|x,x^*) \qquad (14.22.3)$$

for all $x^* \in X^*$, $k \geq k(x)$, and $z \in C$.

Whenever $x \in \mathcal{b}$, $k \geq k(x)$, and $m \in N$, we have

$$P_k(m)x = \sum_{n=0}^{m} \binom{m}{n}[P(c/k)-I]^n x = P(c/k)^m x. \qquad (14.22.4)$$

Since $c = p/q$ and $1/q \in \Delta$, we have

$$\begin{aligned} P_{pk}(qkm)x &= P(1/qk)^{qkm}x = [P(1/qk)^k]^{qm}x \\ &= [P(1/q)]^{qm}\,x \end{aligned} \qquad (14.22.5)$$

for all $k \geq k(x)$ and $m \in N$.

Fix $x^* \in X^*$ and $k,\ell \geq k(x)$. The function

$$f(z) = x^*P_{pk}(qkz)x - x^*P_{p\ell}(q\ell z)x$$

is a Laplace-Stieltjes transform (by (14.22.3)) which vanishes for all $z \in N$ (by (14.22.5)). By Lerch's theorem (cf. [11;Theorem 6.22]), $f(z) = 0$ for all $z \in C$.

Thus

$$P_{pk}(qkz)x = P_{p\ell}(q\ell z)x \quad (z \in C; k,\ell \geq k(x)), \qquad (14.22.6)$$

and therefore $\widetilde{P}(\cdot)x$ is a well-defined entire function.

If $r \in \Delta$ is a positive rational number, we may write $r = m/k$ with $m,k \in N$ and $k \geq k(x)$.

By (14.22.4),

$$\tilde{P}(r)x = P_{pk}(qm)x = P(1/qk)^{qm}x = P(r)x$$

since $qm/qk = r \in \Delta$.

The continuity of $\tilde{P}(\cdot)x$ and $P(\cdot)x$ on Δ (cf. Condition (iii) in the definition of local semigroups) now implies that these functions coincide on Δ.

By (14.22.3),

$$x^*P_{pk}(qkz)x = \int_R e^{-qkzv}\mu_{pk}(dv|x,x^*)$$

$$= \int_R e^{-zu}\mu_{pk}(du/qk|x,x^*) \qquad (14.22.7)$$

$(x^* \in X^*, k \geq k(x), z \in C)$.

The uniqueness property of Laplace transforms and (14.22.6) show that we may define

$$\mu(\delta|x,x^*) = \mu_{pk}(\delta/qk|x,x^*) \qquad (14.22.8)$$

for $\delta \in \mathcal{B}(R)$, $x^* \in X^*$, and $k \geq k(x)$, the right hand side of (14.22.8) being independent of k. By (14.22.1), (14.22.7), and (14.22.8), the proof of the lemma is complete.

14.23 Corollary. If X is reflexive, there exists a unique map $E: \mathcal{B}(R) \to B(Z,X)$ (where Z is normed by $|||\cdot|||$) such that:

(i) for each $x \in Z$, $E(\cdot)x$ is a (strongly) countably additive vector measure with $\mathrm{var}(E(\cdot)x) \leq |||x|||$ (in particular, $||E(\delta)|| \leq 1$ for each $\delta \in \mathcal{B}(R)$); and

(ii) $\tilde{P}(z)x = \int_R e^{-zv}E(dv)x \quad (z \in C; x \in Z)$.

Proof. By Lemma 14.22, the map $x^* \to \mu(\delta|x,x^*)$ is a continuous linear functional on X^*, with norm $\leq |||x|||$ (for each $\delta \in \mathcal{B}(R)$ and $x \in Z$).

Since X is reflexive, there exists a unique vector $E(\delta)x \in X$ such that $x^*E(\delta)x = \mu(\delta|x,x^*)$. By uniqueness, we conclude that E has all the properties stated above.

14.24 Theorem. Let $\mathcal{P} = (P,\mathcal{D},\Delta)$ be a densely defined local semigroup in a reflexive Banach space X, and let $\mathcal{b}$ denote the linear manifold of binomial vectors for $\mathcal{P}$.

Then $\mathcal{P}$ is a spectral local semigroup if and only if $\mathcal{b} = \mathcal{D}$ and the norm $|||\cdot|||$ is equivalent to the given norm $||\cdot||$ on $\mathcal{D}$.

Proof. Sufficiency.

Suppose $\mathcal{b} = \mathcal{D}$ and $|||x||| \le b||x||$ $(b = b(P) < \infty)$ for all $x \in \mathcal{D}$. Let $E: \mathcal{B}(R) \to B(Z,X)$ be as in Corollary 14.23. Then

$$||E(\delta)x|| \le |||x||| \le b||x|| \qquad (\delta \in \mathcal{B}(R), x \in \mathcal{D}).$$

Since $\mathcal{D}$ is dense in X, each $E(\delta)$ has a unique extension as an element of $B(X)$ (also denoted $E(\delta)$), and $||E(\delta)|| \le b$ ($B(X)$-norm!). This uniform boundedness and the strong countable additivity of $E(\cdot)x$ for each $x \in \mathcal{D}$ imply (by density of $\mathcal{D}$) that $E:\mathcal{B}(R) \to B(X)$ is strongly countably additive and var $E \le b$ (cf. Corollary 14.23).

Moreover (since $Z = \mathcal{D}$ in our present situation)

$$\tilde{P}(z)x = \int_R e^{-zv}E(dv)x \qquad (z \in C, x \in \mathcal{D}). \tag{14.24.1}$$

In particular, $E(R)x = \tilde{P}(0)x = P(0)x = x$ for all $x \in \mathcal{D}$ (cf. Lemma 14.22), hence $E(R) = I$ by density of $\mathcal{D}$.

By (14.24.1) and Lemma 14.22,

$$P(t)x = \int_R e^{-tv}E(dv)x \qquad (x \in \mathcal{D}, t \in \Delta) \tag{14.24.2}$$

and

$$\tilde{P}(is)x = \int_R e^{-isv}E(dv)x \qquad (x \in \mathcal{D}, s \in R). \tag{14.24.3}$$

In particular, $||\tilde{P}(is)x|| \le b||x||$ $(x \in \mathcal{D}, s \in R)$, and therefore, again by density of $\mathcal{D}$ in X, $\tilde{P}(is)$ extends uniquely as an element of $B(X)$ (also denoted $\tilde{P}(is)$) with operator norm $\le b$.

We now use the elementary combinatorial identity

$$\binom{z+w}{n} = \sum_{m=0}^{n} \binom{z}{n-m}\binom{w}{m} \qquad (x,w \in C;\ n = 0,1,2,\dots).$$

By Lemma 14.22, if $x \in \mathcal{D}$ and $k \ge k(x)$, we have for all $s,t \in R$ (using absolute convergence of the series involved and the fact $\tilde{P}(is) \in B(X)$):

$$\begin{aligned}
\tilde{P}(is)\tilde{P}(it)x &= \tilde{P}(is)\sum_{m=0}^{\infty}\binom{iqkt}{m}[P(1/qk)-I]^m x \\
&= \sum_{m=0}^{\infty}\binom{iqkt}{m}P_{pk}(iqks)[P(1/qk)-I]^m x \\
&= \sum_{m=0}^{\infty}\binom{iqkt}{m}\sum_{n=m}^{\infty}\binom{iqks}{n-m}[P(1/qk)-I]^{n-m}[P(1/qk)-I]^m x
\end{aligned}$$

$$= \sum_{n=0}^{\infty} \binom{iqk(s+t)}{n} [P(1/qk)-I]^n x$$

$$= P_{pk}(iqk(s+t))x = \tilde{P}(i(s+t))x.$$

By density of $\mathcal{D}$ and the boundedness of the operators $\tilde{P}(is)$, we conclude that $\tilde{P}(i\cdot)$ is a group of bounded operators on R. By (14.24.3) and the group property, we conclude that $E(\delta)E(\sigma) = E(\delta\cap\sigma)$ for all $\delta,\sigma \in \mathcal{B}(R)$ (cf. proof of Theorem 14.11). Thus E is a spectral measure. Let T be the s.t.s. operator with resolution of the identity E. By (14.24.2), $P(t)x = e^{-tT}x$ for all $x \in \mathcal{D}$ and $t \in \Delta$.

For each $x \in \mathcal{D}$, $\tilde{P}(z)x$ is an entire function of exponential type $\leq A_x < \infty$, and it follows that the measures $\mu(\cdot|x,x^*)$ $(x^* \in X^*)$ are supported by the interval $[-A_x,A_x]$ (cf. proof of Lemma 14.22). Hence $E(\cdot)x$ is carried by $[-A_x,A_x]$, and therefore, for $k \in N$ large enough,

$$x = E(R)x = E([-k,k])x \in \mathcal{D}_0,$$

that is $\mathcal{D} \subset \mathcal{D}_0$, and $(P,\mathcal{D},\Delta)$ is a spectral local semigroup.

<u>Necessity</u>. Let $P = (P,\mathcal{D},\Delta)$ be a spectral local semigroup. We have $P(t) = e^{-tT}|\mathcal{D}$ $(t \in \Delta)$, where

$$\mathcal{D} \subset \mathcal{D}_0 = \bigcup_{k=1}^{\infty} E([-k,k])X$$

and E is the resolution of the identity of the s.t.s. operator T.

If $x \in \mathcal{D}$, let $m = m(x)$ be such that $x \in E([-m,m])X$. Then

$$[P(c/k)-I]^n x = \int_{-m}^{m} [e^{-cv/k}-1]^n E(dv)x \qquad (k,n \in N).$$

Since $|e^{-cv/k}-1| \leq e^{cm/k}-1$ for $-m \leq v \leq m$,

$$||[P(c/k)-I]^n x||^{1/n} \leq (e^{cm/k}-1)(\text{var } E\cdot||x||)^{1/n}.$$

Also $\limsup_{n\to\infty} |\binom{z}{n}|^{1/n} \leq 1$ for all $z \in C$, and therefore

$$\limsup_{n\to\infty} ||\binom{z}{n}[P(c/k)-I]^n x||^{1/n} \leq e^{cm/k}-1 < 1$$

for all $z \in C$ and all $k \geq k(x) = [cm(x)/\log 2] + 1$, that is $x \in \mathcal{b}$, and so $\mathcal{b} = \mathcal{D}$.

For each $(\underline{c},\underline{t}) \in \Omega$, $x \in \mathcal{D}$, and $k \geq k(x)$, since $|e^{-cv/k}-1| < 1$ for $-m \leq v \leq m$, we obtain

$$\begin{aligned}
\left\|\sum_{j=1}^{N} c_j P_k(it_j)x\right\| &= \left\|\sum_{j=1}^{N} c_j \sum_{n=0}^{\infty} \binom{it_j}{n} [P(c/k)-I]^n x\right\| \\
&= \left\|\int_{-m}^{m} \sum_{j=1}^{N} c_j \sum_{n=0}^{\infty} \binom{it_j}{n} [e^{-cv/k}-1]^n E(dv)x\right\| \\
&= \left\|\int_{-m}^{m} \sum_{j=1}^{N} c_j [1+(e^{-cv/k}-1)]^{it_j} E(dv)x\right\| \\
&= \left\|\int_{-m}^{m} \sum_{j=1}^{N} c_j \exp(-icvt_j/k) E(dv)x\right\| \\
&\leq \operatorname{var} E \cdot \|x\|.
\end{aligned}$$

Hence $\|P_k(i\cdot)x\|_B \leq \operatorname{var} E\cdot\|x\|$ for all $x \in \mathcal{D}$ and $k \geq k(x)$, that is $|||x||| \leq \operatorname{var} E\cdot\|x\|$ $(x \in \mathcal{D})$, Q.E.D.

Notes and References

§1-2

The approach is based on [19]. See also [4], [9], and [40]. For a generalization of the $H(K)$-o.c. to vector functions, see [17].

Theorem 2.1.1. The proof given here is found essentially in [19] (proof of Theorem 3.4). The construction of Q uses Nagy's argument [41].

§3-4

Theorem 3.1. Cf. [19; Theorem 2.18]. The proof is inspired by an argument of Y. Katznelson (cf. "Sur le calcul symbolique dans quelques algèbres de Banach, Ann. Sci. École Normale Supérieure (3) 76 (1959), 83-123, Théorème 5.1).

Lemma 3.2. Cf. [19; Lemma 2.11]. The details are from [18].

Theorem 4.1. A first version is found in [38]; in the present form, this is the "weak representation theorem" in [30; p. 122]. Theorem 4.3 is [30; Theorem 1].

Proposition 4.4. Cf. [9; Proposition 1].

§5

This section is based on [26], [27], and [29]. Some of the tools are inspired by [47].

Theorem 5.8. Cf. [29; Corollary 4], [17; p. 531], and [1] (for the special case $a \sim b$).

Theorem 5.19. Cf. [29; Theorem 13].

Corollary 5.20 Cf. [26; Theorem 3.2]. See [27] for an elementary direct proof and [24; Theorem 4] for the original special result.

Corollary 5.21. Cf. [26; Theorem 3.6 and Corollary 3.7]). Also [26] for Corollaries 5.22, 5.23, and 5.24.

§6

Theorem 6.8. Cf. [26; Lemma 5.6].

Theorem 6.9. Cf. [26; Theorem 5.7] and [24; Theorem 6]. The proof here uses also an idea from the proof Theorem 3.2 in [37].

Theorem 6.10. Cf. [26; Theorem 5.2] and [23; Theorem 6].

Theorem 6.11. Cf. [26; Theorem 5.8] and [24; Theorem 1].

Theorem 6.13. Cf. [26; Theorem 5.5] and [23; Theorems 8 and 9].

§7

This section is based on [12]. Theorem 7.1 is a version of Mihlin's multiplier theorem.

§8

See [15],[16], and [8].

§9

See [26], [35].

§ 10

Background material (such as Lemma 10.2, 10.4, and 10.5) is taken from [4].

Theorem 10.10. Cf. [26; Theorem 6.2] and [29; Theorem 12].

Corollary 10.12. Cf. [24; Theorem 3].

§ 11

Cf. [37].

§ 12

This section is based on [37],[28], [15], and [35]. Examples are taken from [31] and [35].

Theorem 12.18. Cf. [15; §5]. A more general approach is also contained in that paper.

§ 13

Theorem 13.10. Cf. [19; Theorem 3.13].

Theorem 13.22. Cf. [20; Theorem 2.1], and [21; Theorem 2.1] for the special case $m = 0$.

§ 14

Cf. [22],[30], [33], [34], and [36].

Local semigroups of symmetric operators are discussed in [42].

A general theory of semigroups of unbounded operators is found in [14].

Bibliography

1. C. Apostol, Remarks on the perturbation and a topology of operators, J. Funct. Anal. 2(1968), 395-409.

2. S. Bochner, A theorem on Fourier-Stieltjes integrals, Bull. Amer. Math. Soc. 40(1934), 271-276.

3. C. Caratheodory, Theory of Functions I, Chelsea, New York (1958).

4. I. Colojoara and C. Foias, Theory of Generalized Spectral Operators, Gordon and Breach, New York (1968).

5. N. Dunford and J.T. Schwartz, Linear Operators, Interscience Publishers, New York. Part I (1958); Part II (1963); Part III (1971).

6. R.E. Edwards, Functional Analysis, Holt, Rinehart and Winston, Inc., New York (1965).

7. A. Erdelyi, Higher Transcendental Functions, Vol. I, McGraw-Hill, New York (1953).

8. M.J. Fisher, Imaginary powers of the indefinite integral, Amer. J. Math. 93(1971), 317-328.

9. C. Foias, Une application des distributions vectorielles à la théory spectrale, Bull. Sci. Math. (2) 84(1960), 147-158.

10. A. Friedman, Generalized Functions and Partial Differential Equations, Prentice-Hall, Englewood Cliffs, N.J. (1963).

11. E. Hille and R.S. Phillips, Functional analysis and semigroups, Amer. Math. Soc. Colloquium Publ. 31, Providence, R.I. (1957).

12. L. Hörmander, Estimates for translation invariant operators on L^p spaces, Acta Math. 104(1960), 93-139.

13. L. Hörmander, Linear Partial Differential Operators, Springer-Verlag OHG, Berlin (1963).

14. R.J. Hughes, Semigroups of unbounded linear operators in Banach space, Trans. Amer. Math. Soc. 230(1977), 113-145.

15. R.J. Hughes and S. Kantorovitz, Boundary values of holomorphic semigroups of unbounded operators and similarity of certain perturbations, J. Funct. Anal. 29 (1978), 253-273.

16. G.K. Kalisch, On fractional integrals of pure imaginary order in L^p, Proc. Amer. Math. Soc. 18(1967), 136-139.

17. S. Kantorovitz, Operational calculus in Banach algebras for algebra-valued functions, Trans. Amer. Math. Soc. 110(1964), 519-537.

18. S. Kantorovitz, On the characterization of spectral operators, Trans. Amer. Math. Soc. 111(1964), 152-181.

19. S. Kantorovitz, Classification of operators by means of their operational calculus, Trans. Amer. Math. Soc. 115(1965), 194-224.

20. S. Kantorovitz, A Jordan decomposition for operators in Banach space, Trans. Amer. Math. Soc. 120(1965), 526-550.

21. S. Kantorovitz, The semi-simplicity manifold of arbitrary operators, Trans. Amer. Math. Soc. 123(1966), 241-252.

22. S. Kantorovitz, Local C^n-operational calculus, J. Math. Mech. 17(1967), 181-188.

23. S. Kantorovitz, The C^k-classification of certain operators in L_p, Trans. Amer. Math. Soc. 132(1968), 323-333.

24. S. Kantorovitz, The C^k-classification of certain operators in L_p, II, Trans. Amer. Math. Soc. 146(1969), 61-68.

25. S. Kantorovitz, On the operational calculus for groups of operators, Proc. Amer. Math. Soc. 26(1970), 603-608.

26. S. Kantorovitz, Commutators, C^k-classification, and similarity of operators, Trans. Amer. Math. Soc. 156(1971), 193-218.

27. S. Kantorovitz, Formule exponentielle pour les éléments Volterra d'une algèbre de Banach, C.R. Acad. Sc. Paris 274(1972), 1876-1877.

28. S. Kantorovitz, Commutation de Heisenberg-Volterra et similarité de certaines perturbations, C.R. Acad. Sc. Paris 276(1973), 1501-1504.

29. S. Kantorovitz, Spectral equivalence and Volterra elements, Indiana Univ. Math. J. 22(1973), 951-957.

30. S. Kantorovitz, Characterizations of C^n-operators, Indiana Univ. Math. J. 25(1976), 119-133.

31. S. Kantorovitz, Similarity of certain operators in ℓ^p, Proc. Amer. Math. Soc., 67(1977), 99-104.

32. S. Kantorovitz, Intertwining certain operators in ℓ^p, J. London Math. Soc. (2) 16 (1977), 514-516.

33. S. Kantorovitz, Characterization of unbounded spectral operators with spectrum in a half-line, Comment. Math. Helvetici 56(1981), 163-178.

34. S. Kantorovitz, Spectrality criteria for unbounded operators with real spectrum, Math. Ann. 256(1981), 19-28.

35. S. Kantorovitz and R.J. Hughes, Similarity of certain singular perturbations in Banach space, Proc. London Math. Soc. (3) 42(1981), 362-384.

36. S. Kantorovitz and R.J. Hughes, Spectral representation of local semigroups, Math. Ann. 259(1982), 455-470.

37. S. Kantorovitz and K.J. Pei, Pairs of operators satisfying the Volterra commutation relation, Indiana Univ. Math. J. 23(1974), 1177-1197.

38. S. Kantorovitz and K.J. Pei, Representation of C^n-operators, Proc. Amer. Math. Soc. 48(1975), 152-156.

39. T. Kato, Perturbation theory for linear operators, Springer, New York (1966).

40. F. Maeda, Generalized spectral operators on locally convex spaces, Pac. J. Math. 13(1963), 177-192.

41. B. Sz.-Nagy, On uniformly bounded linear transformations in Hilbert space, Acta Sci. Math. (Szeged) 11(1947), 152-157.

42. A.E. Nussbaum, Spectral representation of certain one-parameter families of symmetric operators in Hilbert space, Trans. Amer. Math. Soc. 152(1970), 419-429.

43. W. Rudin, Real and Complex Analysis, McGraw-Hill, New York (1966).

44. W. Rudin, Fourier analysis on groups, Interscience Publishers, New York (1962).

45. W. Rudin, Functional Analysis, McGraw-Hill, New York (1973).

46. L. Schwartz, Théorie des distributions, Hermann & Cie, Paris (1966).

47. P.H. Vasilescu, Spectral distance of two operators, Rev. Roumaine Math. Pures Appl. 12(1967), 733-736.

48. D.V. Widder, The Laplace transform, Princeton University Press, Princeton.

49. A. Zygmund, Trigonometric Series, 2d ed., Cambridge University Press, New York (1959).

Index